NON-CONVENTIONAL SOURCES OF ENERGY

AF520287

NON-CONVENTIONAL SOURCES OF ENERGY

Copyright © by Laxmi Publications (P) Ltd.
All rights reserved including those of translation into other languages. In accordance with the Copyright (Amendment) Act, 2012, no part of this publication may be reproduced, stored in a retrieval system, or transmitted in any form or by any means, electronic, mechanical, photocopying, recording or otherwise. Any such act or scanning, uploading, and or electronic sharing of any part of this book without the permission of the publisher constitutes unlawful piracy and theft of the copyright holder's intellectual property. If you would like to use material from the book (other than for review purposes), prior written permission must be obtained from the publishers.

Printed and bound in India
Typeset at Goswami Associates, Delhi
First Edition: 2015
UIC-9745-125-NON-CONV. SOURCE ENERGY-MAT
ISBN 978-93-83828-53-1

Limits of Liability/Disclaimer of Warranty: The publisher and the author make no representation or warranties with respect to the accuracy or completeness of the contents of this work and specifically disclaim all warranties. The advice, strategies, and activities contained herein may not be suitable for every situation. In performing activities adult supervision must be sought. Likewise, common sense and care are essential to the conduct of any and all activities, whether described in this book or otherwise. Neither the publisher nor the author shall be liable or assumes any responsibility for any injuries or damages arising herefrom. The fact that an organization or Website if referred to in this work as a citation and/or a potential source of further information does not mean that the author or the publisher endorses the information the organization or Website may provide or recommendations it may make. Further, readers must be aware that the Internet Websites listed in this work may have changed or disappeared between when this work was written and when it is read.

All trademarks, logos or any other mark such as Vibgyor, USP, Amanda, Golden Bells, Firewall Media, Mercury, Trinity, Laxmi appearing in this work are trademarks and intellectual property owned by or licensed to Laxmi Publications, its subsidiaries or affiliates. Notwithstanding this disclaimer, all other names and marks mentioned in this work are the trade names, trademarks or service marks of their respective owners.

Branches

©	Bangalore	080-26 75 69 30	
©	Chennai	044-24 34 47 26,	24 35 95 07
©	Cochin	0484-237 70 04,	405 13 03
©	Guwahati	0361-254 36 69,	251 38 81
©	Hyderabad	040-27 55 53 83,	27 55 53 93
©	Jalandhar	0181-222 12 72	
©	Kolkata	033-22 27 43 84	
©	Lucknow	0522-220 99 16	
©	Mumbai	022-24 91 54 15,	24 92 78 69
©	Ranchi	0651-220 44 64	

PUBLISHED IN INDIA BY

UNIVERSITY SCIENCE PRESS
(An Imprint of Laxmi Publications Pvt. Ltd.)
113, GOLDEN HOUSE, DARYAGANJ,
NEW DELHI - 110002, INDIA
Telephone : 91-11-4353 2500, 4353 2501
Fax : 91-11-2325 2572, 4353 2528
www.laxmipublications.com info@laxmipublications.com

C—
Printed at : Repro Knowledgecast Limited, Thane

Dedicated

To

My Grand Parents

PREFACE

Energy is one of the most important factor for the growth and development of any nation. Actually it is the basic need for human beings *i.e.*, without energy, one can't think about better living. Now-a-days coal, oil and gas are being used as energy sources but due to the increased demand, it is being expected that these resources will loose their existance in near future. In this context, the need for the alternative sources of energy arises so as to meet out the energy needs. These alternative sources may also be knows as "Non-Conventional or Renewable Energy Sources". The present textbook deals with all these renewable energy sources of future. The Book has been divided into nine chapters.

The first chapter deals with energy scenario of World, Asia and India. It also covers certain aspects of transmission and distribution of energy developed in various states of India. The second chapter discusses power majors foray in rural electrification. It covers detail regarding various renewable energy technologies in India. Similarly in third chapter, the scene of power crisis and renewable energy (in India) has been presented. Likewise in chapters 4 to 7, various important renewable energy resources like solar, wind, biomass and fuel cell with related issues, have been presented. Chapter 8 provides brief introductory knowledge of energy obtained from earth, ocean, mini and micro hydel plant, hydrogen and helium-3. In the last chapter 9, issues related with energy conservation, energy audit and cleaner energy have been discussed.

Thus in brief, this book provides good material regarding various non-conventional sources of energy. I hope that this textbook will be proved as a useful book for the students, at the diploma and engineering level. It will also be fruitful for the students, preparing for competitive exams. I would always welcome suggestions from readers/students for the improvement of this book.

—Author

ACKNOWLEDGEMENTS

I appreciate the patience, understanding and support of Mr. Ratnesh Gupta. His valuable comments and suggestions made it possible for me to complete this book on time. I am also thankful for my wife, parents and brothers for supporting me at every stage of preparation of the manuscript for the present book. My thanks are also due to Mr. Saurabh Gupta for bringing out this edition in such a short time period.

—Author

CONTENTS

Chapter **1**

ENERGY

INTRODUCTION TO POWER GENERATION

We all know that energy can neither be created nor be destroyed but can only be converted from one form to another useful form. In reality, energy provides the means of economic growth and social development. For generation of electrical energy, some other forms of energy are needed. Actually energy is related with power where power is the rate of flow of energy and it is normally used in terms of generation and consumption of electricity. Overall it has been proved as an important factor in the progress of mankind.

Electricity (electrical energy and power) has the highest versatility and convenience of use, but statistics show that all of us are facing the shortage of energy. This energy shortage is mainly due to the rapid increment in the population of various countries. For example, in India, population has crossed a dangerous limit, due to which it needs to consume more and more energy for his own. If present trend continues, then in near future our world will be more crowded than that of today.

However, we have some limited sources of power generation (electrical power) such as:

(*a*) Fossil fuels, such as Oil, Coal, Natural gas etc.

(*b*) Hydroelectric (falling water energy)

(*c*) Wind energy (*d*) Solar radiation's energy

(*e*) Geothermal energy (*f*) Biomass energy

(*g*) Ocean tides and waves energy etc.

ENERGY SOURCES AND THEIR CLASSIFICATION

Today, each country draws energy from various (above mentioned) sources. In a broader sense, these can be categorized as commercial and non-commercial sources. The commercial sources use friendly. India has the 5th largest installed power capacity in the world (The first four include-US, Japan, China and Russia-in that order) and is one of the top power consumers. India has an installed power capacity of approximately 150,000 MW, which is 4% of the global capacity. Also the per capita power consumption (per year) is nearly 720 KWh, which is less than 5% of the per capita consumption in USA and less than 20% than that of China. In addition, the global per capita consumption of power is nearly 2340 KWh.

However, few important points, related to Indian power sector include:

- Current installed power generation capacity—150,000 MW (approx.).
- Domination of public sector enterprises—NTPC, NHPC, NPCI and PGCIL.
- Actually 75% share is constituted by thermal power (coal and gas based), 21% by hydroelectricity power, 3.5% by nuclear power and less than 0.5% by wind and solar power.
- Gap between peak electricity demand and production-12%.
- In present situation, nearly 80% of rural areas have been connected to electricity, less than 45% of rural households have access to electricity.

In India, the government has set-up big targets for the past many years but failed every time. During the past three 5-year plans, the achievement rate of target in installing fresh power generation capacity has been less than 50% and the main reasons for this slippage include delays in award of works, financial closure, technology procurements etc. Other reasons may be the law and order problems, lack of trained manpower, different court cases etc. However for future, it is being expected that till 2014, the installed capacity will raise by 78000 MW, especially considering that the current installed capacity is only 150,000 MW. Hence, what is the plan to obtain this massive 50% boost is something that only the government of India can reveal.

However the electricity production in India from various resources like coal, oil, natural gas, hydroelectric and nuclear sources is as given in following figures (Fig. 1.2 to Fig. 1.6).

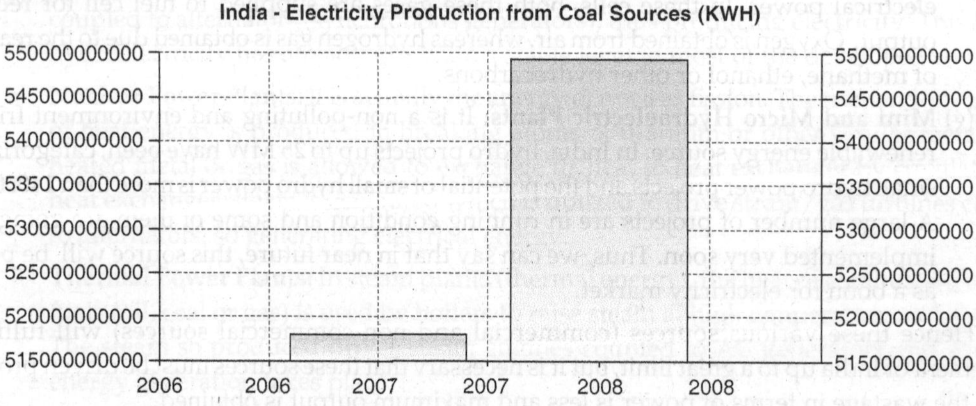

Fig. 1.2: Electricity production from coal sources in India

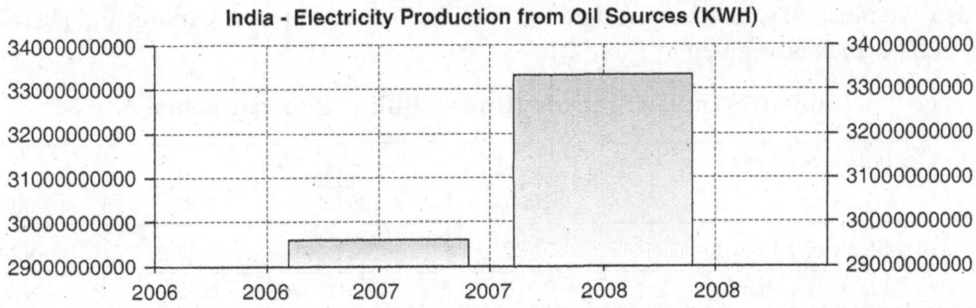

Fig. 1.3: Electricity production from oil sources in India

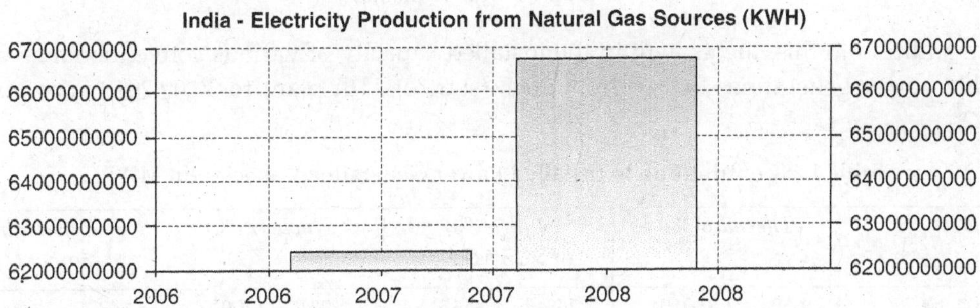

Fig. 1.4: Electricity production from natural gas sources in India

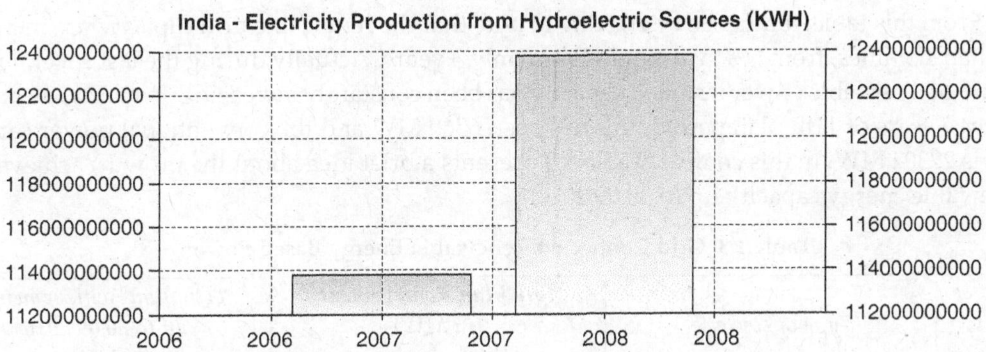

Fig. 1.5: Electricity production from hydroelectric sources in India

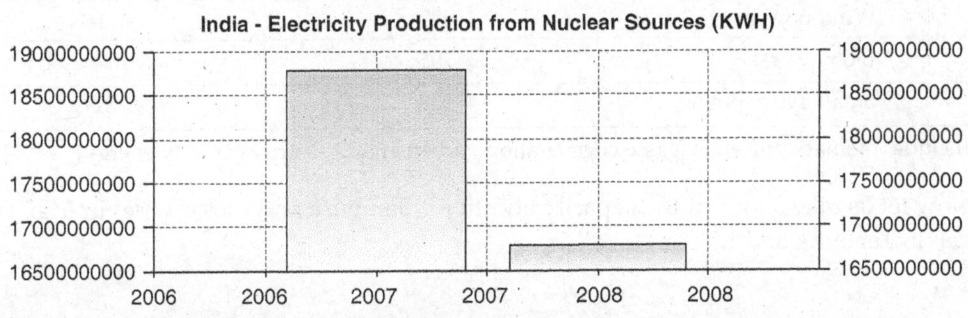

Fig. 1.6: Electricity production from nuclear sources in India

Also, we must take a look at the data for Total Electricity Imports, Exports and Distribution losses in India, which is as given in Table 1.1.

Table 1.1: Indian Scenario (Unit → Billion kilowatt-hours)

S. No.	*Indian Scenario*	*Year*			
		2006	*2007*	*2008*	*2009*
01	Total Electricity Imports	2.957	5.23	9.249	NA
02	Total Electricity Exports	0.216	0.29	0.388	NA
03	Electricity distribution losses	183.043	185.513	193.741	NA

[**Note:** NA = Not Available)

In addition to this, the growth in the installed capacity of various conventional as well as renewable power generation sources from 2007 (start of 11th plan) to 30.09.2010 is as given in Table 1.2.

Table 1.2: Contribution to Installed Power Generation Capacity (in MW)

Date	*Thermal (%)*	*Hydro (%) (> 25 MW)*	*Nuclear (%)*	*Renewable power (%)*
01.04.2007	65% (87015)	26% (34654)	3% (3900)	6% (10258)
.30.09.2010	64% (106518)	22.4% (37328)	2.7% (4560)	10.90% (18155)

From this table, it must be interesting to note that the renewable grid capacity has increased more than 1.5 times, from 6% to around 11% in only 3 years. Actually during the last several years, the share of renewable power installed capacity has been constantly increasing. In reality, during the first, three years of 11th plan period, this has been 6560 MW and the conventional power capacity added is 22302 MW. In this context, Table 1.3 presents a brief idea about the growth/achievement of renewable energy capacity up to 30.09.2010.

Table 1.3: Grid Connected Renewable Energy Based Power

S. No.	*Type of source*	*10th plan Achievement (in MW)*	*11th plan Achievement up to 30.09.2010 (in MW)*
01	Solar power	01	08
02	Wind power	5427	4714
03	Bio power*	795	1079
04	Small hydro power	538	759

[**Note:** *Includes biomass power , bagasse cogeneration , urban and industrial waste to energy]

Now let us take a look at the capacity addition (plan wise and sector wise) in India, which is as given in Table 1.4 and 1.5.

Table 1.4: Capacity Addition: Plan Wise

Plan	*Plan-period*	*Installed capacity (MW)*	*Capacity added (MW)*
Tenth (X)	2002-2007	141,079	36,033
Eleventh (XI)	2007-2012	219,656	78,577

[*Source:* Planning commission, D and B industry research service.]

Table 1.5: Sector Wise Capacity Addition (FY 10 - FY 12) (In MW)

Sector / Project status	*FY 10*	*FY 11*	*FY 12*
Central	10,240	10,301	5,876
Committed	6,490	6,890	3,987
Under construction	3,750	3,411	1,889
Private	4,575	5,561	5,610
Committed	146	726	2,960
Under construction	4,429	4,835	2,650
State	4,710	11,506	3,700
Committed	3,000	7,290	2,540
Under construction	1,710	4,216	1,160

Also the relationship between various renewable resources and targeted as well as projected capacities, with certain assumptions is as given in Table 1.6.

Table 1.6: Projected Contribution of Grid-interactive Renewable Power

(Capacity in MW)

S. No.	*Resource*	*11th plan Target*	*Total Targeted capacity by the end of 11th plan (2012)*	*Total Projected capacity by the end of 12th plan (2017)*	*Total projected capacity by the end of 13th plan (2022)*
01	Solar power	200	200 (approx.)	4000	20000
02	Wind power	9000	16100 (approx.)	27300	38500
03	Biomass power, Bagasse cogen. and waste to energy	1780	3000 (approx.)	5100 (approx.)	7300 (approx.)
04	Small hydro power	1400	3400 (approx.)	5000	6600

Also the Coal and oil demand up to year 2011/12 and other important data for India can be represented in tabular forms as given in Table 1.7 to Table 1.11.

Table 1.7: Coal Demand and Supply Forecasts for India (Million Tonnes)

	2011/12
Demand	872
Domestic Supply	652
Deficit	220

[*Source*: Planning Commission]

Table 1.8: Oil Demand and Supply in India (Million Barrels per day)

Year	*Crude Production*	*Crude Imports*	*Petroleum products demand*	*Self reliance (%)*
2001/02	0.74	1.57	2.10	33
2006/07	0.80	2.20	2.89	26
2011/12	0.90	3.31	4.06	21

[*Source*: Planning Commission]

Table 1.9: Natural Gas Demand Supply in India (Million Cubic Meters per day)

	2001/02	*2006/07*	*2011/12*
Demand	117.8	167.1	216.4
Production	71.2	57.5	43.8
Gap	46.6	109.6	172.6

[*Source*: Planning Commission]

Table 1.10: Demand and Supply Forecasts for Power in India (GW)

	2001/02	*2006/07*	*2011/12*
Peak Load	95.76	130.94	176.65
Installed Capacity	126.04	181.10	242.00
Peaking Capability	88.92	129.82	146.67

[*Source*: CEA]

Table 1.11: Demand for Commercial Energy for Final Consumption (BAU Scenario)

S. No.	*Source*	*Year →* *Units*	*2006-2007*	*2011-2012*
1	Coal	Million tonnes	134.99	173.47
2	Lignite	Million tonnes	16.02	19.70
3	Natural gas	Million cubic meters	18291	20853
4	Oil products	Million tonnes	139.95	196.47
5	Electricity	Billion units	712.67	1067.88

[*Source*: Planning commission]

It has been estimated that in near future:

(*i*) The share of gas based generation will rise to more than 12% in year 2014 (compared to 11% in 1996/1997), because these plants have low capital costs and very low gestation time period.

(*ii*) Likewise the share of oil-based generation will decrease from 7.5% to nearly 5.6%.

(*iii*) The part of coal based generation would remain unchanged due to abundant reserves available.

(*iv*) Also hydro power generation will decrease from 19% in 1996/1997 to approximately 16.9% up to 2014. However, if we use renewable energy sources then it can be said that this generation will increase and would continue to have a very small share (≈ 0.74%) of total generation mix.

Now let us take a look at the data of population (projected) for India, which is as given in Table 1.12.

Table 1.12: Projected Population (Millions) of India, 2007-2025

	History	*Projections*		
Year	*2007*	*2015*	*2020*	*2025*
Population	1165	1294	1367	1431

Hence to meet the higher future demand, India must have various generating sources, so as to increase the generation. Although electric power generation in India, on a commercial basis is almost a century old, but substantial power development efforts began after independence. At the launch of First Five Year plan in 1951, power generation was recognized as a major input for country's economic development and was given high priority. Thus the need for quicker and greater capacity addition, led the Government of India to assume a leading role in large scale power generation and through an amendment of electricity (supply) Act, NTPC and NHPC were set-up in central sector to supplement the efforts of states. Thus NTPC, NHPC and various other sources are giving their valuable services to meet the demand in India. In near future, if we will utilize these sources properly, then there will be no power shortages or power cuts and then India will be on the way of success and prove itself as a developed nation.

Also Energy consumption is an important factor for any country, as it gives the measure of its growth. India ranks 6th in world in total energy consumption and requires to accelerate the development of sector to meet out its growth aspirations. In India, the rising oil import bill has been the focus of serious concerns due to the pressure; it has placed on scarce foreign exchange resources. India is rich in coal and abundantly endowed with renewable energy in the form of solar, wind, hydro and bio energy and has very small hydrocarbon reserves (≈ 0.4% of world's reserve). The distribution of primary commercial energy resources in India is quite skewed. Nearly 70% of total hydro potential is in Northern and Northeastern regions and Eastern regions have approximately 70% of total coal reserves in country. Likewise Southern region has nearly 6% of total coal reserves and 10% of total hydro potential. However for the purpose of comparison, the bar-chart for the energy consumption in United States, China and India is as given in Figure 1.7.

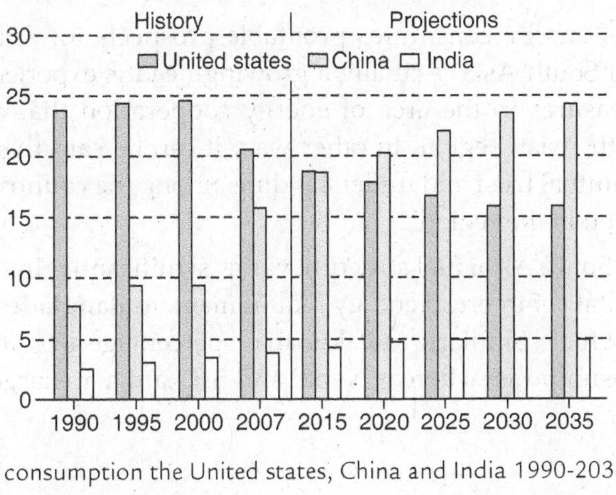

Fig. 1.7: Shares of energy consumption the United states, China and India 1990-2035 (Percent of world total)

ENERGY SCENARIO IN ASIA

Asia is an important region of world hence it becomes necessary to have an idea about the energy scenario in Asia. In this section, we may take the case of southern region. The South Asian region comprises Nepal, Bangladesh, Pakistan, Maldives, India, Bhutan, Sri Lanka and others. The region is currently experiencing a rapid growth in energy demand, so sufficient supply of energy is actually a major challenge, facing economics in the region.

The countries of South Asia (especially **India and Pakistan**) are electrically deficient. **India's** electricity generation capacity is nearly 20% below peak demand. An estimation regarding this shows that nearly 12×10^{10} watts are required to be added to India's capacity over next 25–30 years to enable it to meet the growth of 10% in demand for electricity. Similarly, **Pakistan's** installed generation capacity must increase to nearly 54×10^{9} watts up to year 2018. **Srilanka** is also facing shortage of power and it must improve its production level with an increment above 160 MW. **Bangladesh** must have an installed capacity of 6100 MW. (more than double its current capacity) to meet its peak power demand in future. Situation in Bangladesh is likely to improve, because 3 barge mounted power plants of 100 MW each in Chattagong Khulna and Narayangang are being constructed with recent discovery of gas reserves in Bay of Bengal.

As we all know that both India and Pakistan are facing problem of acute shortage of gas supply. Both these countries have the advantage of bordering two major natural gas rich regions namely the Persian Gulf and central Asia. Actually it is this geographical proximity, which can give the basis for mutually beneficial cooperation between both these countries. It is clear that South Asian countries would have ever increasing energy demands and they actually lack the resources and capital to meet this demand. According to World Bank report, the major problem to its economic development in coming years is the bad quality of energy infrastructure of South Asia. Other issue regarding this is, that South Asia lies in the vicinity of countries with rich natural gas and other resources. Given the economical attraction of pipeline transportation of oil and gas, this becomes the most preferred means for import of natural gas into region from the neighboring countries. Thus a strong case is made for coordinated action to meet challenges and exploit the potential in energy sector in that region.

The major field of energy constitutes profitable prospects for strengthening cooperation between the countries of South Asia. Actually a growing need is experienced for initiating some confidence building measures in the area of energy cooperation that could lead to extensive cooperation among South Asian region. In other way, it can be said that area of cooperation of energy would enhance mutual trust and understanding among the countries and provide basis for more economic integration in the region.

With reference to South Asia, fuel structure varies significantly. In **India,** coal constitutes as much as nearly 30% of total commercial energy requirement. In **Bangladesh**, oil (53%) and natural gas (≈ 47%) are dominant fuels. Similarly in **Pakistan** oil percentage 46% and natural gas has ≈ 38% share of total energy consumption, whereas **Nepal and Sri Lanka** are largely dependent on oil.

ENERGY SCENARIO IN WORLD

We know that the population of the world is continuously increasing and it is being expected that in future (up to 2025), it will cross the limit of 7900 million. (In this context projected capacities for electricity generation will be discussed later on). Hence in this relation, the population data for world, its various regions as well as countries is as given in Table 1.13 to Table 1.15.

Table 1.13: Projected Population (Millions) of World, 2007–2025

	History	*Projections*		
Year	*2007*	*2015*	*2020*	*2025*
Population	6650	7256	7610	7932

Table 1.14: Projected Population (Millions) of Various Regions, 2007–2025

		History	*Projections*		
S. No.	*Region ↓ Year →*	*2007*	*2015*	*2020*	*2025*
01	North America	441	477	500	523
02	Central and South America	464	511	538	563
03	Middle East	199	231	251	270
04	Africa	939	1098	1197	1297

Table 1.15: Projected Population (Millions) of Various Countries, 2007–2025

		History	*Projections*		
S. No.	*Country ↓ Year →*	*2007*	*2015*	*2020*	*2025*
01	United States	302	327	343	359
02	Russia	142	138	135	132
03	China	1321	1386	1421	1443
04	Japan	128	126	124	122

Also the data for world scenario including electricity imports , exports and distribution losses in various regions and countries is as given in Table 1.16 to Table 1.22 and the essential figures for world primary energy demand and world marketed energy use by fuel- type is as drawn in Figure 1.8 and Figure 1.9.

Table 1.16: World Scenario (Unit → Billion Kilowatt-hours)

S. No.	*World Scenario ↓ Year →*	*2006*	*2007*	*2008*	*2009*
01	Total Electricity Imports	614.9656	625.3404	612.9166	NA
02	Total Electricity Exports	616.3835	628.8509	614.0989	NA
03	Electricity distribution losses	1622.622	1652.402	1657.252	NA

[**Note:** NA = Not Available]

Table 1.17: Total Electricity Imports in Various Regions (Billion Kilowatt-hours)

S. No.	*Region ↓ Year →*	*2006*	*2007*	*2008*	*2009*
01	North America	66.96264	71.53929	81.104	70.54072
02	Central and South America	55.85023	54.79787	54.469	NA
03	Europe	368.186	374.11	354.329	NA
04	Middle East	8.338	7.565	7.494	NA
05	Africa	35.7038	36.8695	37.33072	NA
06	Asia & Oceania	26.269	27.525	30.6229	NA

[**Note:** NA = Not Available]

Table 1.18: Total Electricity Exports in Various Regions (Billion Kilowatt-hours)

S. No.	*Region ↓ Year →*	*2006*	*2007*	*2008*	*2009*
01	North America	66.96264	71.53929	81.104	70.54072
02	Central and South America	54.81467	52.53603	52.447	NA
03	Europe	359.588	365.864	344.127	NA
04	Middle East	4.632	4.777	8.139	NA
05	Africa	31.3887	30.7003	29.08737	NA
06	Asia & Oceania	26.895	30.38	31.0215	NA

[**Note:** NA = Not Available]

Table 1.19: Electricity Distribution Losses in Various Regions (Billion Kilowatt-hours)

S. No.	*Region ↓ Year →*	*2006*	*2007*	*2008*	*2009*
01	North America	357.2773	362.6113	339.4838	NA
02	Central and South America	158.2482	163.1746	167.7647	NA
03	Europe	252.8698	255.4292	259.0398	NA
04	Middle East	80.56068	87.35344	95.735	NA
05	Africa	67.18454	65.70251	69.2132	NA
06	Asia & Oceania	541.7618	554.8244	560.615	NA

[**Note:** NA = Not Available]

Table 1.20: Total Electricity Imports in Various Countries (Billion Kilowatt-hours)

S. No.	*Country ↓ Year →*	*2006*	*2007*	*2008*	*2009*
01	France	8.522	10.782	10.683	19.213
02	Germany	48.464	45.953	41.67	41.853
03	China	5.389	4.251	3.842	6.006
04	Russia	5.115	5.67	3.105	3.0659
05	United States	42.69131	51.3957	57.02055	52.42954
06	United Kingdom	10.282	8.613	12.294	6.609

Table 1.21: Total Electricity Exports in Various Countries (Billion Kilowatt-hours)

S. No.	Country ↓ Year →	2006	2007	2008	2009
01	France	71.863	67.595	58.689	44.913
02	Germany	65.441	62.508	61.77	54.132
03	China	12.271	14.566	16.644	17.386
04	Russia	20.927	18.468	20.738	17.7
05	United States	24.27134	20.14359	24.08345	18.11118
06	United Kingdom	2.765	3.398	1.272	3.748

Table 1.22: Electricity Distribution Losses in Various Countries (Billion Kilowatt-hours)

S. No.	Country ↓ Year →	2006	2007	2008	2009
01	France	31.805	31.613	32.916	NA
02	Germany	28.804	29.518	30.118	NA
03	China	181.67	196.255	191.829	NA
04	Russia	107.589	104.855	109.24	NA
05	United States	266.2768	264.1826	245.8815	245.9446
06	United Kingdom	28.456	27.453	28.195	NA

[**Note:** NA = Not Available]

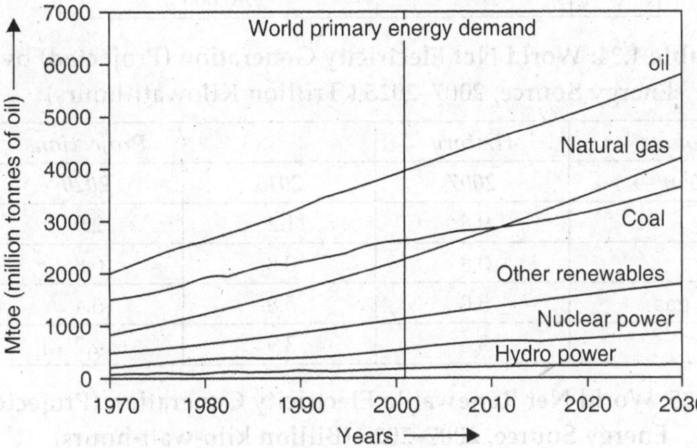

[*Source*: International energy agency]

Fig. 1.8: Future projections for world primary energy demand

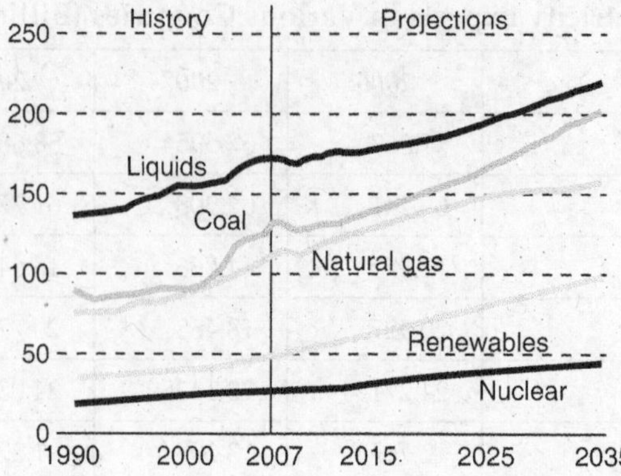

Fig. 1.9: World marketed energy use by fuel-type 1990-2035 (Quadrillion BTU)

The world total installed capacity and net projected electricity generation (With fuel resources) is as given in Table 1.23 and 1.24 respectively. As we know that now a days, it becomes imperative to look at renewable generation, so in this respect, the data for world net renewable electricity generation is as given in Table 1.25.

Table 1.23: World Total Installed Capacity (MW)

S. No.	*Year*	*Installed Capacity (MW)*
1	2007	93.927
2	2008	120.903
3	2009	159.766
4	2010	196.630
5	2011	240.000

Table 1.24: World Net Electricity Generation (Projected) by Energy Source, 2007–2025 (Trillion Kilowatt-hours)

S. No.	*Resource ↓*	*History*	*Projections*		
	Year →	*2007*	*2015*	*2020*	*2025*
01	Coal	9.8	11.2	12.9	15.0
02	Liquids	0.8	0.8	0.8	0.8
03	Natural gas	5.0	5.8	6.4	6.8
04	Nuclear	3.6	3.9	4.2	4.5

Table 1.25: World Net Renewable Electricity Generation (Projected) by Energy Source, 2007–2025 (Billion kilo-watt-hours)

S. No.	*Renewable Energy Source*	*History*	*Projections*		
	Year →	*2007*	*2015*	*2020*	*2025*
01	Solar	6	95	126	140
02	Wind	165	682	902	1115
03	Hydro power	2999	3689	4166	4591
04	Geothermal	57	98	108	119
05	Other	235	394	515	653

Also the energy consumption (projected) of the world (with fuel sources) and its various regions is as given in Table 1.26 to Table 1.29.

Table 1.26: Energy Consumption (Projected) of the World, 2007–2025 (Quadrillion Btu)

	History		*Projections*	
Year →	*2007*	*2015*	*2020*	*2025*
	495.2	543.5	590.5	638.7

Further, Table 1.26 (as discussed above) may be extended on the basis of various resources. The data are as given in Table 1.27.

Table 1.27: Energy Consumption (Projected) of the World by considering Different Fuel Resources (Quadrillion Btu)

S. No.	*Resources ↓*	*History*		*Projections*	
	Year →	*2007*	*2015*	*2020*	*2025*
01	Coal	132.4	139.1	152.4	167.8
02	Liquids	174.7	179.3	186.0	197.2
03	Natural gas	112.1	129.1	141.2	150.2
04	Nuclear	27.1	32.2	37.4	41.1
05	Other	48.8	63.8	73.4	82.4
	Total	**495.2**	**543.5**	**590.5**	**638.7**

Also the data for energy consumption (projected) in various regions of the world (2007–2025) is as given in Table 1.28.

Table 1.28: Energy Consumption (Projected) in Various Regions of the World, 2007–2025 (Quadrillion Btu)

S. No.	*Region ↓*	*History*		*Projections*	
	Year →	*2007*	*2015*	*2020*	*2025*
01	North America	123.7	124.3	129.4	134.9
02	Central and South America	28.0	32.1	35.5	38.7
03	Middle East	25.1	32.9	36.5	39.1
04	Africa	17.8	20.8	22.5	24.6

However this table may also be discussed in detail by considering various resources such as coal, liquids, natural gas etc. and can be represented as in Table 1.29.

Table 1.29: Energy Consumption (Projected) in Various Regions of the World by Considering Different Resources, 2007–2025 (Quadrillion Btu)

(*a*) North America

S. No.	*Source ↓*	*History*		*Projections*	
	Year →	*2007*	*2015*	*2020*	*2025*
01	Coal	24.6	23.8	24.4	25.0
02	Liquids	49.4	47.5	48.1	49.4
03	Natural gas	29.2	28.3	30.1	32.2
04	Nuclear	9.6	10.1	10.8	10.9
05	Other	11.0	14.6	16.0	17.4
	Total	**123.7**	**124.3**	**129.4**	**134.9**

(*b*) Central and South America

S. No.	Source ↓ / Year →	History 2007	Projections 2015	2020	2025
01	Coal	0.9	0.9	1.0	1.2
02	Liquids	12.2	13.4	13.6	14.4
03	Natural gas	4.9	6.0	7.4	8.0
04	Nuclear	0.2	0.3	0.4	0.5
05	Other	9.8	11.4	13.0	14.6
	Total	**28**	**32.1**	**35.5**	**38.7**

(*c*) Midele East

S. No.	Source ↓ / Year →	History 2007	Projections 2015	2020	2025
01	Coal	0.4	0.4	0.3	0.3
02	Liquids	13.3	15.0	16.1	17.7
03	Natural gas	11.2	17.0	19.3	20.2
04	Nuclear	0.0	0.1	0.2	0.3
05	Other	0.3	0.4	0.5	0.5
	Total	**25.1**	**32.9**	**36.5**	**39.1**

(*d*) Africa

S. No.	Source ↓ / Year →	History 2007	Projections 2015	2020	2025
01	Coal	4.2	4.2	4.3	4.7
02	Liquids	6.4	7.2	7.4	8.0
03	Natural gas	3.3	5.1	6.1	6.9
04	Nuclear	0.1	0.2	0.2	0.2
05	Other	3.7	4.2	4.5	4.9
	Total	**17.8**	**20.8**	**22.5**	**24.6**

[**Note:** Totals may not equal to the sum of components due to independent rounding]

In addition to this, world net electricity projected consumption (2010–2025) and the scenario of energy revolution for world may also be easily represented in more convenient manner as given in Figure 1.10 and Table 1.30.

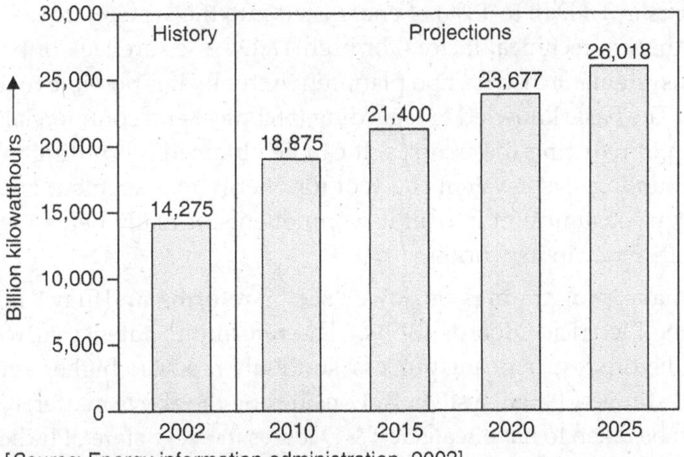

[*Source*: Energy information administration, 2002]

Fig. 1.10: World Net Electricity Consumption, 2010–2025

Table 1.30: Projected Scenario of Energy revolution [In GW] (2020–2050)

S. No.	*Year →* / *Resource*	*2020*	*2030*	*2040*	*2050*
1	Coal	859	623	517	466
2	Oil / Diesel	204	105	34	7
3	Gas	1490	1818	1923	1929
4	Nuclear	175	9	0	0
5	Photovoltaic	199	728	1330	2033
6	Wind	950	1834	2242	2731
7	Bioenergy	211	306	408	505
8	Geothermal	41	70	107	140
9	Hydropower	994	1091	1183	1257
10	Ocean energy	14	28	46	63

TRANSMISSION AND DISTRIBUTION LOSSES

Introduction

We have talked about power generation as well as consumption of India, but this description would not be worthwhile, until we don't talk about the losses occurring during the channel (*i.e.*, from generation to consumption). Thus, it is required to discuss about the losses, called as "Transmission and Distribution losses". In India, average T&D losses are nearly 23% of electricity generated, but according to studies carried out by TERI (*i.e.*, independent agency), these losses have become as high as 50% in some states of India. State regulatory commission provides accurate estimation of transmission and distribution losses as the losses level directly affects the sales and power purchase requirements and so has a bearing on determination of electricity tariff of a utility by the commission.

Technical losses in T&D (8 to 10%) of energy fed into the system are normal. Actually some major reasons contributing technical factors for high T&D losses are lack of proper investment in T&D systems and unsystematic distribution planning. Actually it is possible to reduce losses within short period of time. The basic knowledge regarding (technical and commercial losses) losses is the first step in direction of reducing these losses. It can be obtained by putting in place a system for accurate energy accounting. This system is a tool for energy management and helps in breaking down the total energy consumption into all its components. It really helps the utility in bringing accountability and efficiency in its working.

Some states have programmes of power sector reforms and have taken some steps to restructure their State Electricity Boards (SEBs.) The reforming states, that were reporting T&D losses of nearly 20% before restructuring process, suddenly reported higher amount of losses after carrying out detailed studies of their system. By considering the above matter, it becomes necessary that some steps must be taken to have realistic T& D losses in every state of India and also to reduce them up to a minimum limit. Some solutions of above problems may be:

(*a*) Schemes must be made for giving incentive awards to those utilities, which are able to minimize these transmission and distribution losses beyond a certain fixed limit.

(*b*) Consumers must be made aware about high penalties on the unauthorized use of electricity, through publicity campaigns.

(*c*) For taking remedial measures to reduce T & D losses to a minimum value and the financial institutions must be encouraged to provide easy loans.

Thus overall it can be said that above mentioned steps may be taken to have real assessment of transmission and distribution losses occurring in various states of our country.

Some important issues:

(*i*) Status of metering and steps needed for early installation of same is very important issue. In reality, the feasibility of obtaining the required targets (goals) may be an important point for consideration regarding financial assistance being provided by central government for installation of meters.

(*ii*) Mitigating power theft is another important issue. The electrical power theft is a problem faced by all electric utilities. Due to the theft, loss of revenue takes place. Power quality of electricity obtained becomes also poor, so voltage dips and low voltages are produced. Thus it must be an issue for discussion.

(*iii*) Last but not the least point is the implementation of energy audit schemes. It must be obligatory for all industries and utilities to carry out energy audits of their systems.

Components of T & D losses:

In the process of supplying electricity to consumers, energy losses occur and are called as technical and commercial losses.

(*a*) **Technical losses:** These losses occur due to energy dissipated in conductors and equipments used for transmission, transformation, sub-transmission and distribution of power. These losses can be further subdivided as:

(*i*) Transmission losses

(*ii*) Sub-transmission losses

(*iii*) Distribution losses

Where voltages for various losses (above mentioned) are as shown below:

(*i*) Transmission losses → 400 KV/220 KV/132 KV/66 KV

(*ii*) Sub-transmission losses → 33 KV/ 11 KV

(*iii*) Distribution losses → 11 KV/0.4 KV

(*b*) **Commercial losses** → These losses occur in electrical power system network due to following reasons:

(*i*) Estimating unmetered supply of energy

(*ii*) Defective meters

(*iii*) Error in meter reading

(*iv*) Due to pilferage

In detail, the losses in different elements of T&D system are as given in Table 1.31.

Table 1.31: Losses in Different Elements of T&D System

	**System element*	*% Power losses*	
		Minimum	*Maximum*
(*a*)	Step up transformers and EHV transmission system	0.5	1.0
(*b*)	Transformation to intermediate voltage level, Transmission system and step down to sub-transmission voltage level	1.5	3.0
(*c*)	Sub-transmission system and step down to distribution voltage level	2.0	4.5
(*d*)	Distribution lines and service connections	3.0	7.0
(*e*)	**Total losses**	**7.0**	**15.5**

[**Source*: SEBs]

One of the important point regarding T&D losses is that transmission and distribution losses in most developed countries are very less (nearly 9 to 10%), whereas in our country, it is much more than that.

Reasons for high T&D losses: (technical and non-commercial)

The major reasons for these losses can be categorized as follows:

(*i*) **Technical losses:** Some factors contributing high technical losses are:

(*a*) Since there are too many stages for transformations of voltage, so these losses occur.

(*b*) Improper reactive compensation and inadequate management of load also increases above losses.

(*c*) One important reason is haphazards growth of sub-transmission and distribution system with short-term objective of extension of power supply to new areas.

(*d*) Since in agricultural pumping (in rural areas), in coolers, air conditioners and industrial loads in urban areas, bad quality of equipments as well as instruments are being used, which affects the efficiency of power system network.

(*e*) Improper investment on transmission and distribution (i.e. low investment) results in overloading of distribution network without augmentation.

Similarly, some factors contributing high commercial losses are as given below.

(*ii*) **Commercial losses:** Major reason for T&D losses (commercial) is theft and pilferage. These two factors *i.e.* theft/pilferage of energy is committed by two categories of consumers called as:

(*i*) Non-consumer and

(*ii*) Bonafide consumer

- Some consumers (non-consumer) avail unrecorded supply by tapping the conductors (low tension feeder).
- Some of the bonafide consumers commit the pilferage by damaging the measuring equipments, installed at their premises, which provides an important factor contributing T&D losses.
- Some methods by which these consumers have illegal consumption of electricity are:

(*i*) Changing sequence of terminal wiring.

(*ii*) Errors produced in meter reading and recording.

(*iii*) Stopping meters by remote control method.

(*iv*) By placement of powerful magnets or disturbing the rotation of meter disc.

(*v*) By making extensions of loads (unauthorized).

(*vi*) By meter burnings, etc.

Chapter 2 RURAL ELECTRIFICATION

INTRODUCTION

Rural electrification is a programme for socio-economic development of rural areas. The main objectives are to trigger economic developments and generate employment by providing electricity for productive uses in agricultural and rural industries and improve the quality of life of rural people by supplying electricity for lighting of homes, shops and public places in villages.

In reality, rural electrification in India made great progress in 1980s and more than 200,000 villages were electrified first time. In Bihar, Rajasthan, Uttar Pradesh, West Bengal and Orissa, maximum villages were without electricity, so to reduce this problem one important solution was only the electrification of these villages by applying any of the possible solutions. Actually rural electricity uses supply of energy for two types of programmes.

(*a*) Production oriented activities like minor irrigation rural industries etc.

(*b*) Electrification of villages

⇓

In a proper way, a village is considered as a "**electrified village**" if electricity is being used within its revenue area for any purpose whatsoever. It can also be said that a village will be deemed to be electrified if electricity is used in the inhabited locality within the revenue boundary of village for any purpose whatsoever. These (above mentioned) definitions of village electrification were prepared after consulting with state governments as well as state electricity boards. In a broader sense, a new proposed coverage of village electrification is as given below:

(*a*) The ratings of distribution transformers and low tension lines to be provided in the village, would be finalized as per the anticipated number of connections, decided by consulting with Zila Parishad/District administration, who will also issue the necessary certificate of village electrification on completion of work.

(*b*) All panchayat offices, community centers (Public places) avail supply of power on demand.

(*c*) The number of electrified households may be nearly 10% (as minimum value), before any village is declared electrified.

- **Power majors foray in rural electrification:** For taking up the electrification of villages and households of country, the state owned "**Rural Electrification Corporation of India Limited (RECIL),** which have made agreements with "**POWER GRID"** and **"NHPC"**. These corporations such as "Power grid", "NTPC" and "NHPC", which cover more than 1/3rd of India's electricity generation and transmission, have been persuaded by "RECIL" to electrify the remaining villages and rural households of India, under a massive programme launched by Government of India, called as "**Accelerated Rural Electrification Programme"** (AREP). Under this programme, the RECIL will act as a disbursing agency for the Government grants/loans.

(*a*) Now, let us take a look at RECIL. This Corporation was incorporated as a company under Companies Act, 1956 in 1969 and its main objective was to use financing rural electrification schemes in our country. The present mission is to act as a client friendly and development oriented organization for financing and promoting projects, covering power generation, consumption, transmission and distribution network in our India. The other goals, to be achieved by this organization are:

(*i*) To mobilize funds from various sources including raising of funds from domestic and international agencies and sanction loans to SEBs, state governments power utilities, RECs, non-government organizations (NGO) and private developers.

(*ii*) To improve the various kind of activities like use of new renewable energy sources, transmission and distribution system, maintenance and modernization, for optimization of reliability of supply of power to rural as well as urban areas both including desert, hilly areas as well as remote areas.

(*b*) Similarly, a scheme such as "**Rajiv Gandhi Grameen Vidhyutikaran Yojana"** would be implemented through REC. Using this scheme; nearly 89% capital subsidy would be provided for projects, such as:

- Decentralized distributed generation (DDG) and supply system from conventional sources of power generation for habitations and villages, where ministry of non-conventional energy sources (MNES) would not be providing electricity supply.
- Provision of power as well as distribution transformers of appropriate KVA rating capacity in each habitation and village.
- Creation of Village Electricity Infrastructure (VEI) for electrification of all unelectrified villages and habitations.

(*c*) Likewise "**Kutir Jyoti Programme"**, which provides, financing of electrification of all unelectrified [below poverty line (BPL)] households in our India by 100% subsidy. Prior Commitment of states would be taken before sanctioning these projects (for those projects to be eligible for capital subsidy under this scheme) for:

- Provision of requisite revenue subsidy to state utilities as needed. (Under the Electricity act 2003).
- Deployment of Franchisees for management of rural distribution in projects financed under this scheme.

Actually it is the responsibility of state governments to make adequate arrangements for supply of power to rural areas. It must be ensured by the states that there would be no discrimination in the supply hours between rural and urban households.

Now, let us take a look at the present status (approx. Values) of un-electrified villages in India.

Table 2.1: Status of Un-electrified Villages in India

S. No.	*States/U.T.*	*Un-electrified Villages*
1	Andhra Pradesh	0
2	Arunanchal Pradesh	1668
3	Assam	5383
4	Bihar	18395
5	Delhi	0
6	Jharkhand	20235
7	Goa	0
8	Gujarat	73
9	Haryana	0
10	Himachal Pradesh	312
11	Jammu and Kashmir	113
12	Karnataka	355
13	Kerala	0
14	Madhya Pradesh	1904
15	Chhattisgarh	867
16	Maharashtra	5018
17	Manipur	361
18	Meghalaya	2354
19	Mizoram	137
20	Nagaland	455
21	Orissa	20994
22	Punjab	0
23	Rajasthan	12595
24	Sikkim	25
25	Tamil Nadu	0
26	Tripura	367
27	Uttar Pradesh	11492
28	Uttaranchal	548
29	West Bengal	1553
30	A & N Island	175
31	Chandigarh	0
32	D & N Haveli	0
33	Daman & Diu	0
34	Lakshadweep	0
35	Pondicherry	0
	Total All India:	**105379**

[*Source*: Progress report published by CEA: http://power min.gov.in].

- **Electrification of Tribal Villages (Poor homes) of India:** Rural electrification in India has suffered very badly in last decade due to poor financial health and operational features of state electricity boards. Although more than 85% of total villages have been electrified over the years and a number of villages are needed to be electrified yet, but the people in rural areas cannot afford the electricity cost and they meet their needs through firewood, cowdung, kerosene and agricultural residue and due to inefficient exploitation of these sources of power, environmental degradation has become a greater factor. In reality, an action plan on 100% village electrification has been prepared, in which rural electrification would be treated as a basic minimum service under the Prime Minister Gramodaya Yojna (PMGY).

In India, more than 15000 villages are in remote areas and they can only be electrified by using non-conventional sources, because these villages cannot be easily connected through conventional grid systems. Biomass and photovoltaic based power plants can be set up in villages so as to meet the energy demand without harming the atmosphere. Local institutions like "**panchayats**" might play an important role in the implementation maintenance and operation of such power plants. It will really fulfill the energy needs up to some extent.

- **New Approach towards Electrification:** Although these poor homes of villages has no light facility but by using some basic schemes, these homes can be electrified. In this case, no new transmission lines and no new distribution lines will be required except the existing ones. Similarly, no house wiring will be required. In view of this, CFL technology has become as a blessing for this scheme. CFL tube will be complete with reflector, small inverter inside and other small auxiliaries, with battery arrangement. The set of these CFL tubes and battery costs will be very less, nearly equal to 1% expense of annual budget of Government of India.

Thus all poor people will have good light in their homes at a cost of nearly ₹ 1 per day. There will be no formality and no harassment, but the only requirement is that electricity must be available for battery charging only. One solution is that in each village, when there is no grid connection, at least a few people will have electricity in their home and out of them only few homes can install battery charger and provide service of charging batteries. In some areas, if wind flow is easily available, then by using wind turbines, batteries can be charged to electrify those villages and it will cost very less compared to installation and operation of transmission and distribution lines, so if above mentioned scheme is implemented, then obviously it will be like a heaven on the earth, for the poor people of villages.

These villages can be electrified also by using either PV or biomass technologies. Nearly 5000 of these villages were proposed to be electrified during 10th five year plan, using Photovoltaic and Biomass technology. In today's scenario, some villages in Bhopalpatnam and Usoor blocks in Dante wade district (in Chhattisgarh state) are going to be electrified by solar photovoltaic systems for all households and community uses. The project was supported by MNES (Ministry of non-conventional energy sources) and MTA (Ministry of Tribal Affairs), Government of India. This project is being completed through CEL (Central Electronics Limited), Sahibabad, which is actually involved in manufacturing and marketing of solar photovoltaic products and systems. Under this project, training of local manpower for repair and maintenance of systems at village level is being done for ensuring long term satisfactory performance of systems. However, similarly the process of production of Biomass to produce electrical power supply can be implemented in the village itself. Through biomass gassifier plant fed on locally available wood, electricity can be provided to residents. Now we can take a basic case to show the future target using biomass to produce electricity so as to electrify the rural areas/villages (Fig. 2.1).

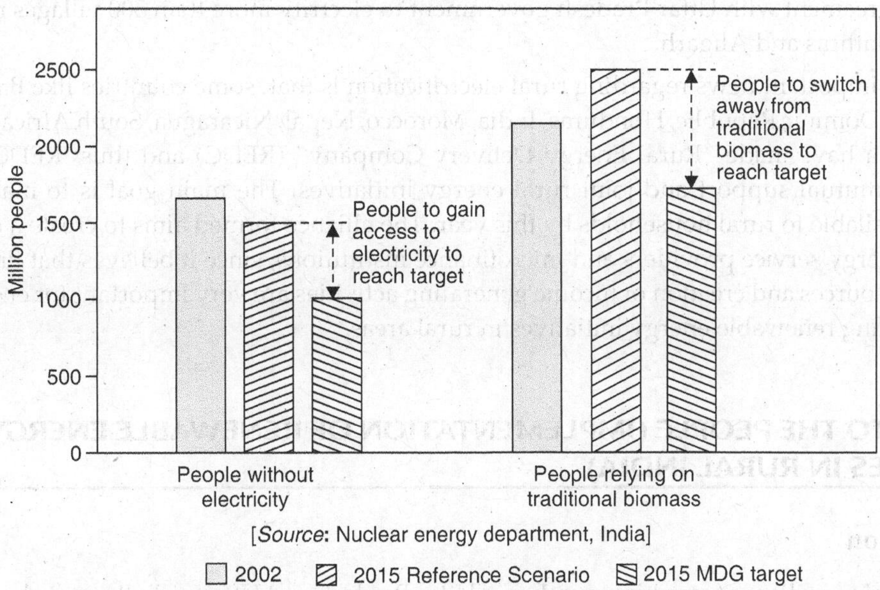

Fig. 2.1: Future target using biomass to produce electricity to electrify the rural areas/villages.

These above mentioned projects regarding biomass and PV systems give more benefits for tribal population. One of the greatest advantages is that these remote rural areas, which had never seen electricity, will be now able to enjoy the benefits of technology. It can also be said that overall improvement in the quality of life will be a superb case for rural population.

CONTRIBUTION OF PUBLIC SECTOR TOWARDS RURAL ELECTRIFICATION

Power grid decided to take a plunge in this area to help rural people. At present "**POWER GRID**" is implementing various distribution network improvement schemes in Bihar, Mizoram, Meghalaya and Goa states. The work is carried out at a very fast rate for rural electrification of more than 2000 villages in Vaishali and Muzaffarpur districts of Bihar state. The project being implemented by **POWER GRID** is unique in such a way that for the first time, integrated planning has been made for RE works, which includes supply backbone strengthening along with latest high voltage distribution and providing service connections. The ministry of power, Govt. of India has decided to bring in other PSUs (Public Sector Units) to take up challenge regarding village electrification in our India, after taking the initiative by "**Power grid**" in this area. Similarly, NTPC and REL (Reliance Energy Limited) have started projects for rural electrification and these projects will be completed using locally available renewable energy sources. NTPC has commissioned two pilot distributed generation (DG) projects, one at Jemara village in Korba and other in Jarha-Chetwa village in Sonebhadra,

Uttar Pradesh. The capacity of plant of Jarha-Chetwa is approximately 11900 watts, using nearly 170 solar photovoltaic of 70 watts each and one 10 KVA power conditioning unit. NTPC is also planning to have electrification projects in the vicinity of its existing power stations and is in the process of identifying unelectrified villages in Angul (Orissa state). Likewise, Reliance Energy Limited has made an agreement with Uttar Pradesh government to electrify more than 500 villages near Agra, Mathura, Hathras and Aligarh.

One important news regarding rural electrification is that, some countries like Bangladesh, Bolivia, the Dominic Republic, Honduras, India, Morocco, Nepal, Nicaragua, South Africa, Tanzania and Uganda have made "Rural Energy Delivery Company" (REDC) and thus 'REDC' alliance stands for mutual support and joint rural energy initiatives. The main goal is to make energy services available to rural households by this year. The alliance formed aims to create a direct link between energy service providers and microfinance institutions, since it believes that financing of renewable sources and creation of income generating activities are very important for encouraging and sustaining renewable energy initiatives in rural areas.

POWER TO THE PEOPLE (IMPLEMENTATION OF RENEWABLE ENERGY TECHNOLOGIES IN RURAL INDIA)

Introduction

In India, various villages (very high number in Uttar Pradesh and Uttaranchal) await electrification and a number of households that possess actual connections, continue to endure erratic and poor quality power supply. RETs (renewable energy technologies) hold the promise of, and potential for, bridging rural India's demand-supply gap in energy in general and electricity in particular. In those areas, where grids are not available, RETs are relevant for those remote and rural areas. Due to some problems created in publicity, allocation of and access to financial resources and technology adaptation to local needs, India's approach to RET dissemination and popularization has had limited success. These problems can be reduced by utilizing alternative approaches of a decentralized nature. One of the alternative has been provided by TERI (The Energy and Research Institute) with its project called as **"Uttam Urja Project."** The project was initiated with support from ICEF (India-Canada Environment Facility) and is now being consolidated and expanded under REEEP (Renewable Energy and Energy Efficiency Partnership) programme, which is being coordinated by British high commission. The project actually provides energy services through entrepreneurs of Tehri-Garhwal and Dehradun districts of Uttaranchal and Bikaner district of Rajasthan. The important work is to develop an energy service network in corporating local NGOs, dealers and retailers.

In this scheme, by paying a specified amount or cost, he may have the ownership of the device (such as domestic lightening system, kisan torch, solar mixie etc.). To meet the demand of rural areas, the "**Uttam Urja Project Team**" has developed and produced large domestic lightening systems of rating 50 watts (peak). Similarly small variants of DLS (30 WP) and a lantern (3 watts) have also been developed to provide facility to those rural households, who have never seen electricity. Thus, the low cost (*i.e.*, cost effective–reduced by 50%) products, such as solar mixies, DLS systems, solar water heaters, in collaboration with private manufacturers, have modernized the life of village people. Local assembly of all systems is undertaken in Bikaner and Rishikesh, considerably reducing the system cost and the time of delivery, while enhancing the efficiency after sales service. In Rajasthan state (Bikaner and Rishikesh), for financing above systems, TERI has made tie-ups with SHGs (self help groups) and banks. Now, let us take a look at cost comparison chart of various devices, such as DLS, torch, solar panels etc.

Table 2.2: (Cost Comparison)

S. No.	*Product*	*Cost of Assembly (Rupees)*	*Selling Prices (Rupees)*
1.	Large domestic lightening systems (50 WP)	12500	12500
2.	Standard domestic lightening systems (35 WP)	8698	8700
3.	Solar panels (5 WP)	864	1150
4.	Pyrolysers	18750	18750

Regarding awareness generation, a lot of efforts are being made. By using wall painting banners, maintenance manuals, handbooks and newspapers, slogans explaining RETs are being promoted. These efforts are being given proper shape by the demonstration in offices, government concerns, schools, markets, service centers, video and music shows and in door to door promotions. Non-government organizations are also playing important role by promoting these through meetings with farmers, school teachers, panchayat samities and health workers. Nowadays women are forming an important interest group in RET dissemination. The project training modules encourage women to understand the working of various RET-based household projects. For providing help to women, the project has developed entrepreneurship for solar systems and initiated a revolving fund.

In villages, middle and high income groups are able to purchase domestic lightening systems, but low income groups cannot afford it. To overcome this situation, solar panels and other devices have been developed specially for low-income groups. However, if we consider the financing options such as NGOs, banks etc., then these two can be considered as suitable financing routes. Similarly, **"Kisan Credit Cards"** a government initiative-may be another financing avenue. Likewise national banks issue cards, that allocate credit to farmers in proportion to the land mortgaged and provides same interest rate as that of banks. Thus overall "**Uttam Urja Project"** is very successful, with a large number of products, which are continuously being used in organizations, schools and healthcentres.

ADVANTAGES OF TERI UTTAM URJA PRODUCTS

However, there are many advantages of above projects, but some of them may be listed below for better understanding. These include:

(*i*) Quality of life

(*ii*) Benefits to children and women

(*iii*) Entrepreneurship development

(*iv*) Environmental gains

(*i*) Quality of Life

Due to these projects certainly the life of rural households has been improved. Solar fans, solar power television sets and emergency lights are being proved as a gift to these villages. In the days of summer (mostly in the month of June-July), solar fans provide respite from the sweltering temperatures. Also the danger of animal attacks has been automatically reduced, because they do not approach well-lit-houses.

(*ii*) Benefits to Children and Women

The pressure of daily work in the schedule (*i.e.* in the light of sun) has been reduced. The involvement of some women in after-sales service is an example of women's empowerment. Due to the use of these systems (environmentally friendly), the quality of air is improved which is actually advantageous for women and children.

(*iii*) Entrepreneurship Development

This project provides a strong basis for entrepreneurship development or in other sense, it may be said that it not only provides employment, but is also successful in inspiring entrepreneurship. Through the creation of a market for RETs, local capacity has gained more strength with the external world.

(*iv*) Environmental Gains

One important benefit is to get air free from pollution. Since by using these devices in (VUP) rural households, there is no bad effect on environment, (*i.e.* reduced air pollution and also reduction in green house gas emissions) so the enhanced lightening is much more tangible.

Thus in brief, we can say that in **"Uttam Urja Project"**, the commercial dissemination of renewable energy technologies is possible. It helps to work with well-established local organizations in initiating the project and gaining trust. Last but not the least point is that a policy shift from subsidizing products to boosting finance and awareness is imperative.

FULL RURAL ELECTRIFICATION BY 2012

In India, Planning Commission's approach to 10th Five Year Plan (2002—2007) fails to give suggestions on rural electrification. It's working groups (report to power-ministry) has said that it is necessary to define minimum number of rural households, to be given power connections before a village is declared as electrified village. In reality, government official must do a survey in collaboration with Panchayati Raj institutes. Although Power Ministry considered year 2007 of complete rural electrification, but commission's report consider year 2012 as a more proper year for full electrification of rural households. According to its report, grid connections will not be easy for these villages, so some systems like solar lanterns and domestic lights must be used, by making a fund with contributions from beneficiaries to run.

One special benefit is that a number of unemployed rural youth will be employed through these schemes/ projects. As according to recent report, power generation in the rural villages may be possible using Biomass, small hydropower projects, wind energy, solar energy etc. (*i.e.* renewable energy technologies). For a bright future of these villages, decentralized systems, involving village communities in planning, technology and collecting revenue, is actually required *i.e.* for full rural electrification as early as possible. The rural electrification programme must be taken up on the basis of 90% grant from the government and only 10% contribution from users. Thus by overall efforts from government, NGOs, self-help groups and village persons, it will be possible to have a good quality of life in rural villages very soon.

Chapter 3

CRISIS OF POWER IN INDIA AND RENEWABLE SOURCES

INTRODUCTION

The shortage of power in India is really a big issue. Most of the states in our country are suffering from power crisis. In recent times Maharashtra has been experiencing a severe power crisis. Most of the parts of Maharashtra are facing power cut situation for nearly six hours a day (average), because there is no addition of any power plant for so many years. After a long time 1700 MW Enron–Dhabol project was in the hands of Maharashtra government after nearly a decade. Even the Dhabol plant returns to its operation but the distress in this state will not completely disappear. Actually changes in the power sector industry took place in 1990s and it was suggested by the central government to have future growth of power generation with private sector funding. After that most of the projects took place in generation, transmission and distribution through centrally owned companies such as Power Grid Corporation, National Thermal Power Corporation, National Hydro Power Corporation etc. and state electricity boards. In 1990s, there was a boom of symposia and seminars by the Independent Power Producers Association (IPPA) to educate government about how the structure must be changed to make it IPP-friendly. We all know that the growth in power sector generation over the past decade is very less and the starting of the private sector Enron project, has been proved as a real disaster for the country. Most of the power professionals warned the country about the glaring infirmities in the project, but politicians sold the project to country. We can hope regarding this matter, that the asset already created will be put to use and public will have to take a big beating in the process.

One important weakness of industries is non-viability of tariff. In year 2001–2002, the supply cost was approximately ₹ 3.50 per unit and the realization was only ₹ 2.40. Other weakness in Indian situation is under investment in transmission and distribution system. Political parties made it an issue and political decisions continued to deploy the resources. Actually transmission and distribution losses is very high because of the additionally extended supply to rural consumers, results in provision of small amount of energy over long distances (leading to high T&D losses). Similarly Pilferage and Theft are other issues, which contribute to transmission as well as distribution losses.

The important thing is that generation plants must run at high capacity factors with low running costs. Transmission and distribution losses must be made minimum and all energy supplied must be metered and paid properly. In this context, proper evaluation of coal, natural gas, LNG (alternative fuels) must be made for escalation factors over time, while deciding on new generation capacity.

However in brief, it can be said that in India, the electricity requirements have grown up to a great limit and the demand has been running ahead of supply. Electricity generation and transmission progress in India is very inefficient compared to developed countries. Generating capacity in our country is used on an average for 3600 hours, but in Japan it is being used for more than 6000 hours in a year. Likewise T&D losses in India is very high consisting of both technical as well as commercial losses. The generation is very less in our country due to following reasons:

(*a*) Due to erratic Monsoons.

(*b*) Delay in delivery of equipments.

(*c*) Lack of coordination in construction activities.

(*d*) Interstate political disputes.

(*e*) Building material's shortage.

(*f*) Non-availability of drives and devices useful in generation plant.

(*g*) Shortage of good qualities of raw material etc.

Thus it can be said that some factors such as sharp increase in demand, bad uses of equipments already installed in power generating plants, high Transmission and distribution losses, shortage of coal are actually responsible for power crisis in our country.

NEED FOR RENEWABLE RESOURCES

"Promoting freedom from fossil fuels ...... but the time is running out ...... soon there will be nothing left to burn on the earth."

There is actually a red alert regarding the fuels used for power generation. The fossil fuels are completely depleting and most probably after some years or some decades, there will be no more fuel. [Actually this information is as according to Indian Renewable Energy Development Agency (IREDA).] In power sector, all India installed capacity of electrical power generating stations under utilities was 104917.5 MW on 31st March 2002 consisting of 26261.22 MW hydro, 74428.82 MW thermal, 2720 MW nuclear and 1507.46 MW wind. However India's plan for electric power (for zero deficit power) by 2012 is as given in Table 3.1.

Table 3.1: India's Plan for Electric Power (For Zero Deficit Power) by 2012

	Thermal (Coal) (MW)	*Gas/LNG/ Diesel (MW)*	*Nuclear (MW)*	*Hydro (MW)*
Installed capacity as on March 2001	61,157	Gas: 10,153 Diesel: 864	2720	25,116
Additional capacity (2001–2012)	53,333	20,408	9380	32,673
Total capacity as on March 2012	115,794 (53.0%)	31,425 (14.6%)	12,100 (5.6%)	57,789 (26.8%)

[*Source:* Tenth and Eleventh five year plan projections]

From the data, it is clear that these shortages will be increasing day by day as GDP growth increases to nearly 10%. Under these situations, power sector must increase its existing capacity (nearly double) up to 2012 so as to meet the higher growth trajectory and also accomplish the targeted mission of ***"power to all"*** by 2012. It is only possible, when renewable sources such as solar, wind, geothermal etc. would be used without any pollution to environment and human beings. To study this case, let's take an example of ***Sweden***, which is set to be the first country, in the world to make a transition from an oil based economy to one based on renewable. In reality, the purpose of Swedish government is to replace all the fossil fuels with renewables and so counter the twin threats of climate change and dwindling oil reserves. Sweden is doing concentrated efforts to reduce its dependence on fossil fuel and to use renewable energy source. Nowadays it has only 32% of energy, being produced by oil. For example, the heating requirements of this country are being carried out by geothermal energy. It also expects to boost the use of bio fuels and expand other types of renewable sources like wind power, wave power etc. Similarly for the (paper and fuel industries) industries, biomass is being used to generate electric power. Thus it can easily be hoped, that Sweden will be poised to become oil free by 2020. Likewise in our country, if renewable sources will be utilized to generate electrical power, then our country may stand in the first row of developed nations in future. So it shows the actual importance of the fact that why we need renewable sources and also tells us that in future every country must use renewable sources instead of fossil fuels as early as possible for the growth of country. In this relation, we must have a look at grid connected renewable energy sources and annual energy generations in India (See Table 3.2).

Table 3.2: Annual Energy Generation from Renewable Sources (Billion KWh)

S. No.	*Source*	*Annual energy generation in billion KWh*
1	Solar photovoltaic	350.4
2	Wind (Onshore)	219
3	Biomass	88.72
4	Bagasse	26.3
5	Waste to energy	26.28
6	Geothermal	70.1
7	Small hydro	46

[*Source*: WISE, January 2011]

FOSSIL FUEL HAZARDS

In present scenario, world is using mostly coal, oil and gas instead of renewable sources (*i.e.*, nearly 85% of total energy). The fossil fuel combustion is actually the main cause of carbon dioxide emissions, which is growing with fast rate per year. The green house gases from the production of the energy are affecting the atmosphere badly. Actually most of the "GLOBAL WARMING" observed over the last 50 years is attributable to human activities and due to this climate change would result if 21st century needs were met without a major reduction in carbon emissions of global energy system.

However, if these green-house gas emissions as well as carbon dioxide emissions are not stopped, then it will affect our environment badly and its effect on the human bodies will be really dangerous. Main pollutants emitted in the combustion of fossil fuels are sulphur and nitrogen oxides, carbon mono-oxide and suspended particulate matter. The environmental impact of a host of energy are linked emissions–including suspended fine particles and precursors of ozone and acid deposition–contribute to local and regional air pollution and ecosystem degradation and so "energy insecurity" is just not a myth, but a red alert against which the various development agencies have to work forward.

NON-CONVENTIONAL ENERGY – A WAY TO ENERGY GAIN

Energy is the basic source of all the activities of human beings. The demand of energy in world is continuously increasing and projections point out towards a doubling of global energy demands in the decade starting 2020. We all know that at present fossil fuel is the major source of energy generation, but the global sources of fossil fuel will have become dearer due to their depletion. So one day will come, when more than 55% of cumulated energy worldwide would be met through renewable sources. [Growing economics, especially of Asia has been gifted with sufficient resource base]. Non-conventional energy technologies are consistent both for grid linked energy generation and transmission. Now technology should be pursued right from this moment and be informed of the barriers in technology applications of renewable sources.

In the year 1970s, it was felt that renewable energy sources can be used for sustainable energy growths and provide a role of helper for human beings. During previous years, this source has been lifted to new heights, due to the support from government. The activities in various sectors cover a number of R.E.S. such as solar energy, wind energy, small hydro power, biogas, biomass and other emerging technologies. Small hydel power generation, which is used specially for hilly regions and remote areas, has a potential of more than 500 MW. Likewise the power generation from wind power takes place in India and it is rated as 4th rank with wind power capacity. Similarly wood stoves and biogas plants (quantity in millions) are also in use in today's scenario. Solar energy is also being utilized as solar photovoltaic and solar lanterns etc. Our country possesses world's largest assemblage of solar PVs, consisting of about 500,000 PV systems. State governments and utilities provide remunerative power purchase agreements and arrangements for wheeling, banking and buy back of power for renewable sources of energy. The Indian Renewable Energy Development Agency (IREDA) is only agency of its kind in the world, dedicated to financing of renewable energy products.

Efforts are needed to make these sources popular among people. People of India should be informed properly about these sources. The fact that, these alternatives are environment friendly and easily replenishable, must be told to the human beings through awareness programmes. The Ministry of Non-conventional energy has strong determination to have nearly 24000 MW from Non-Conventional and Renewable energy by the year 2012. Therefore the actual requirement in today's scenario is to have sufficient policies with a goal to provide impetus through streamlining the structure of Renewable and Non-Conventional energy. In this case one important fact is that we have very good plans to get our goals, but we are not so much enthusiastic to meet the required level of optimization of our potential. However if there is a strict regiment by which renewable and non-conventional energy sources are utilized, India is sure to have adequate measure of success.

RENEWABLE ENERGY AND ITS RESOURCES

Nowadays renewable energy is popular due to its manifold advantages. Technology to utilize the forces of nature for doing work to supply human needs is as old as the first sailing ship. Actually there is a fundamental attractiveness about harnessing such forces in an age which is very conscious of environmental effects of burning fossil fuels. All the renewable sources, such as wind, waves, sun, tides and the heat from radioactive decay in earth's mantle as well as biomass are all abundant and ongoing, hence the term "renewables". The Ministry of Non-Conventional Energy (MNES) observes the 20th August—the birthday of Late Rajiv Gandhi, the former Prime Minister of India, as "**Rajiv Gandhi Akshay Urja Diwas**" (Renewable Energy Day) as he had a vision of taking the country into 21st century through application of modern science and technology. India is to create awareness about the use of renewable energy technologies for various applications. For creating awareness regarding renewable energy, an Education park has been set up in Kolkata. West Bengal Renewable Energy Development Agency (WBREDA) has set up this park in an area of nearly 1.5 acres. The main purpose of establishing this park is to create awareness and give publicity amongst people in general and children in particular about the use and benefits of renewable energy technologies. This park focuses on the working of various renewable energy technologies and demonstrates models of solar PV systems, micro hydel power plants, tidal plant as well as biomass based gasifier plant.

All the renewable energy resources can be summarized as follows:

(*i*) **Solar energy:** Solar energy is one of the types of energy, which is obtained from sun. Solar electric generation has a very low capacity factor. Also the incoming radiations have a very low intensity so efficiency of harnessing sun's energy is very less. One of the most important method of converting sun's energy to electricity is by using photovoltaic cells. It has the application in remote areas, telecommunication equipments, satellites, signaling equipment, lanterns and many more areas. Of course these PVs are increasing strongly as their efficiency is increasing and the cost is continuously decreasing. But the cost per unit of electricity—at least 10 times that of conventional sources, limits its potential to supplementary applications on buildings where its maximum supply coincides with peak demand.

Sometimes for a stand alone system, collected energy is stored during sunshine period and utilized during darkness or cloudy days. Although the cost of PVs is low but the efficiency is very low, so research continues into ways to make the solar collecting cells less expensive and more efficient. Likewise a solar thermal plant has a system of mirrors to concentrate the sunlight on to an absorber, the energy then being used to drive turbines. The concentrator is like a mirror, which tracks sun's path and Absorber is a device located at the focal point and converts solar radiation to heat (nearly 400°C), which is transferred into a fluid such as synthetic oil. Finally fluid is used to drive a conventional turbine and generator. A number of such types are being operated in various regions of the world.

Thus the solar energy is providing main role for future electricity generation. In future, the main application of it will be of direct heating. The need of heat, up to nearly 60°C is in hot water systems and for 60° – 110°C in industry level. Together these may account for a significant proportion of primary energy use in industrialized nations. Finally it can be said that in future, nearly 10% of total primary energy in various countries may be supplied by direct solar thermal techniques and to some extent this will substitute for base load electrical energy.

(*ii*) **Wind energy:** Wind energy is again an important source of power generation. In this case, the output of power is proportional to cube of the wind speed *i.e.*, if wind speed is doubled then output will be nearly 8 times compared to previous value. In today's scenario only few areas are there by which power can be harnessed but many areas are required to be found so as to have large amount of power generation. One important fact is that consumers depending on wind power, need alternative power sources to cope with calmer periods. For generation of electrical energy, wind turbines are required to be used, which actually have a high steel tower to mount the generator and also have rotors with three blades up to 50 m long. The foundation of this system needs hundreds of cubic meters of reinforced concrete. These wind mills with turbine systems and other auxiliaries are set up only in those areas, where wind-flow is easily available.

Most economical size of wind turbines is now up to 2 MW electricity, grouped into wind farms up to 200 MW electricity. Most of these turbines operate at about 25% load factor over the course of a year (European average) but some reach up to nearly 33%. An estimation has been done by a survey that in near future in the year 2030, wind will be providing more than 5% of total world electricity.

(*iii*) **Biomass energy:** It is another means to generate power. Growing crops of wood or other kinds are used to burn as a fuel for the purpose of electricity generation. However the generation using biomass is limited, but now slowly and slowly, awareness regarding this valuable source of energy is increasing day by day. It has been estimated that by the year 2030, biomass fuelled electricity production will be triple that of today's value.

(*iv*) **Geothermal energy:** In this kind of source, hot steam (underground steam) from the earth surface is utilized for production of electricity. These geothermal resources lie in New Zealand, USA, Japan, Italy, Philippines etc. For example, in New Zealand nearly 420 MW electricity produces over 7% of electricity. It is estimated that the power output will be tripled by 2030.

(*v*) **River energy:** In this source, the energy of falling water is utilized or in other sense it can be said that the potential energy of falling water is utilized to generate electrical power. One main benefit of hydro systems is their capacity to handle seasonal (as well as daily) high peak loads.

(*vi*) **Tide energy:** When there is a large tidal range, then by utilizing the tides, turbines can be driven and electricity can be generated. However, placing free standing turbines in major coastal, tidal current appears to have much greater potential and it is being explored.

It has been estimated that currents with velocities of 2 to 3 meters per second are ideal and the kinetic energy involved is equivalent to a wind speed of over 200 km/hr. It means that 1 MW electricity tidal turbine rotor would be less than 20 meter diameter, compared with 60 m for a 1 MW electricity wind turbine. Further progress and improvements are taking place in this field and these are expected to be commercial very soon. As according to estimation, it will increase to nearly 3 billion in 2030, including wave power.

(*vii*) **Waves energy:** However if the power from waves are harnessed, then we may have much more energy compared to tidal energy. Generally this energy is being used as a useful source in UK. Air displaced by waves in a hollow concrete structure would create electrical energy through the coupled generators.

RENEWABLE ENERGY SCENE IN INDIA

As per the latest research, a great portion of India's population still lives in rural areas. Renewable energy projects such as solar, wind and hydropower based electricity are the main factors to provide these rural areas with energy, where power is in short supply. These renewable technologies will not only meet the growing energy demands but also will keep the environment clean. In other sense, it can be said that it is an effective option for meeting a society's ever increasing demands, which provides a degree of a national energy security.

India is the only country, that has a separate ministry for non-conventional energy sources. To provide attention to these factors (as mentioned above), DNES *i.e.,* Department of non-conventional energy sources was created. After sometime, it was renamed as MNES (Ministry of Non-conventional energy sources) to increase the deployment of nearly commercial or already commercial renewable energy technologies. To have greater involvement of the private sector, the MNES was restructured on the basis of end use applications of technologies through horizontal integration of various technologies. One of the important source is solar electricity and our country is turning to the use of solar energy in various villages. Similarly wind generated electrical energy is other factor of India's strategy to boost the use of renewable energies. The growth of this wind energy source in India is actually due to an effort to provide incentives to wind power consumers. Likewise hydro power projects are giving shape to India's energy sector. During the last decade, a great effort has been done for use in different sectors. Now a days, India is in a position today to offer 'state-of-the-art' technology to other developing countries and is poised to play a leading role in the global movement towards sustainable energy development. India has world's largest programmes for renewable energy. Ministry (as discussed above) is also involved in the implementation of these programmes for development and utilization of these renewable energy based technologies such as solar photovoltaic, wind power generation, biomass combustion, mini and micro hydro power, etc. Also it continuously takes care of various newly emerging technologies, such as chemical sources of energy, fuel cells etc. Thus in reality, efforts at war level are being done, which are providing fruitful results to human being including rural village people.

Chapter 4

SOLAR ENERGY

INTRODUCTION

Solar energy is the energy obtained from sun. We use sun for drying our clothes as well as for cooking but recently we are able to produce electricity. If sun's power is harnessed properly, then it will be enough to meet all our power needs. Solar energy is available in abundance and at no cost. The generation of electrical energy from solar energy is of special importance for India due to the following reasons.

(*a*) There exists a need for supply of electrical energy in remote areas. For conservation of fossil fuel resources such as gas, oil, coal, it is also required.

(*b*) It is also necessary to reduce the atmospheric pollution and so in this sense solar cells are of great importance.

WAYS OF HARNESSING SUN'S ENERGY

The sun's energy can be used in three main ways, as given below:

(*i*) **Solar cells (Photovoltaic cells):** It is also called as photovoltaics that convert light energy into electrical energy. In the sunshine period, enough power can be harnessed from sun to run light bulbs and other auxiliaries. Actually it was developed to provide electricity to satellites, but now a days it is being used in watches, calculators and many other devices.

(*ii*) **Solar water heating:** It is used in those areas, where sun's radiations heat the water in glass panels. Water can be pumped through the pipes in the panel. Pipes are painted black and thus absorb more heat from sun and get hot. It helps out our central heating and reduces our electricity bills also.

(*iii*) **Solar Furnaces:** In this case a huge array of mirrors is used to concentrate sun's energy and so a high temperature is produced, which is utilized for furnace work.

Out of these three cases, as mentioned above, solar cells are of great importance. It has been proved as a good source for renewable energy throughout world. A latest report about this renewable energy source tells us that leading Japanese auto manufacturer company, Honda Motor is venturing into the renewable energy market by mass producing next generation solar panels for household and industrial use. Basically all these solar cells would be made up of non-silicon compound materials and generates electricity with less CO_2 during its production compared to conventional sources.

GENERATION OF SOLAR CELL

At present four generations of solar cells exist, which are called as first generation, second generation, third generation and fourth generation solar cells (Fig. 4.1).

(*a*) **First generation solar cells:** These types of cell cover a larger amount of surface area. These cells are single layer P-N Junction diode or in other way it can be said that these cells are typically made using a silicon wafer (also called as silicon wafer based solar cells). It is actually the dominant technology in the commercial production of solar cells, accounting for more than 80% of solar cell market.

(*b*) **Second generation solar cells:** The second generation of photovoltaic materials is based on the use of thin film deposits of semiconductors. In order to reduce the material's cost (intrinsic to Ist generation PV technology) several thin film technologies are being pursued to enable higher volumes to be produced at a lower cost. Different technologies under investigation are amorphous silicon, polycrystalline silicon, microcrystalline silicon, cadmium telluride, copper indium selenide/sulfide etc.

(*c*) **Third generation solar cells:** These types are very different from other two cells. It is a semiconductor device, which do not rely on a traditional *p-n* junction to separate photo generated charge carriers. The types are called as photo electrochemical cells, polymer solar cells and nanocrystal solar cells.

(*d*) **Fourth generation solar cells:** In these cells, a multispectrum layer is made using composite photovoltaic technology with polymers and nanoparticles. Then the thin multispectrum layers can be stacked to make multispectrum solar cells, more efficient and cheaper based on polymer solar cell and multijunction technology.

HOW DO PHOTOVOLTAICS WORK?

The word photovoltaic is a combination of Greek word for light and the name of the physicist Allesandr Volta. This photovoltaic is the direct conversion of sunlight into electricity. In reality, the process of conversion is based on the photoelectric effect (discovered by Alexander Becquerel in 1839). Some materials exhibit a property called as photoelectric effect and so photons of light are

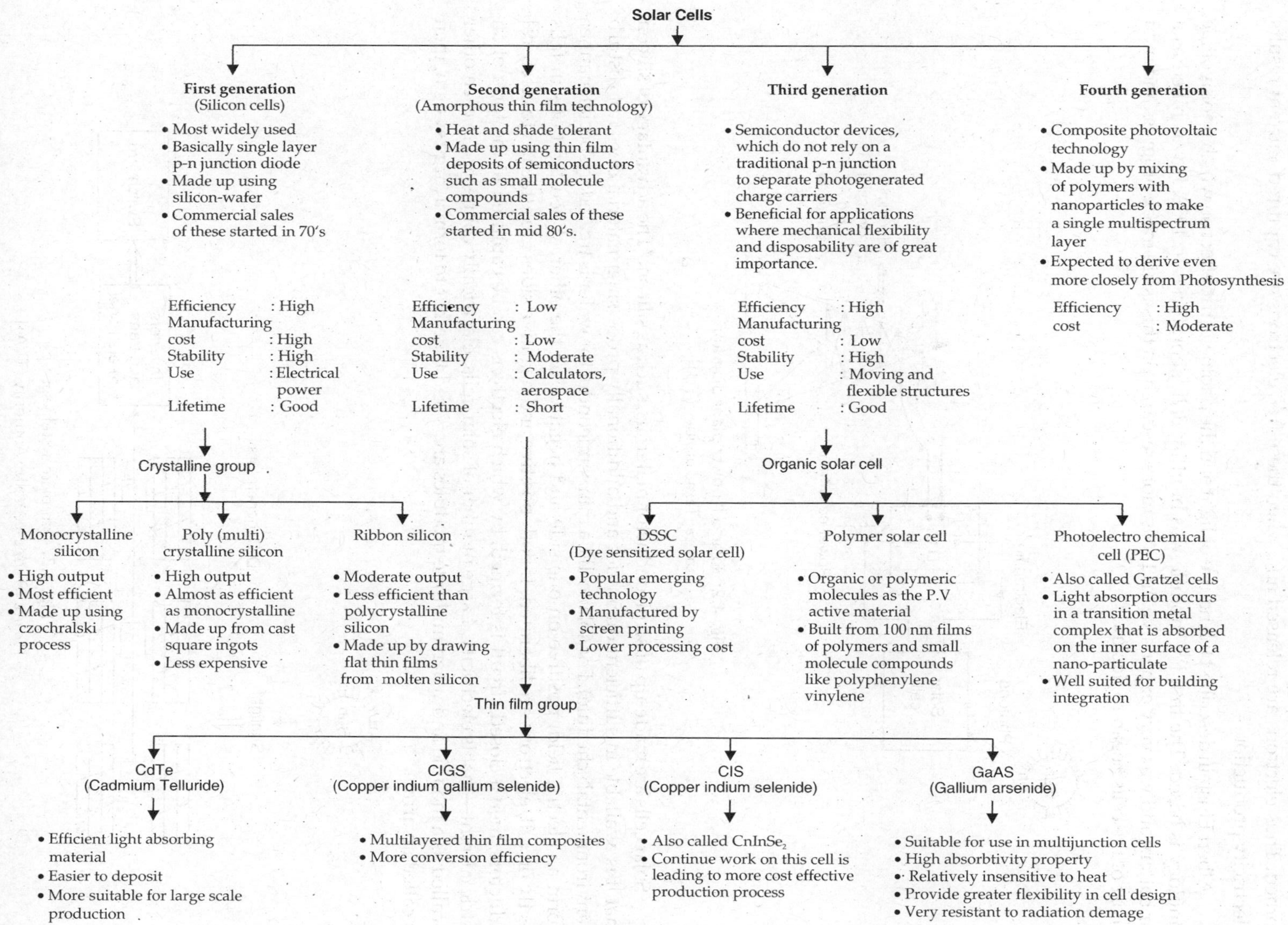

Fig. 4.1: Classification of solar cell

absorbed. Thus electrons are released and when these free electrons are captured, current results in electricity production.

Albert Einstein described the nature of light and the photoelectric effect on which photovoltaic technology is based. The first PV module was built by Bell laboratories in 1954. It was billed as a solar battery and was very expensive. In 1960s more research on this issue made progress and then the cost of the cells began to decline.

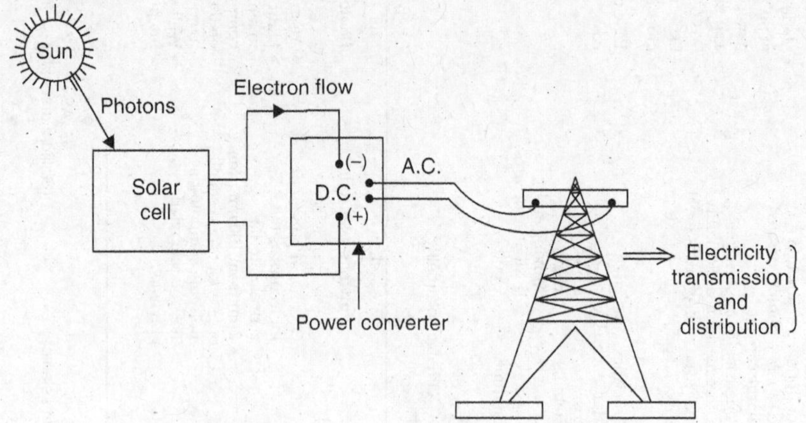

Fig. 4.2: Solar cell power plant diagram
[*Source*: Texas utilities (TU)]

Solar cells are made-up of semiconductor materials, such as silicon. (The advantage of silicon is that it is available in sufficient quantities and additionally processing material does not disturb the environmental schedule). For solar cells, a thin semiconductor wafer is treated in special manner to form an electric field, positive on one side and negative on the other. When light of sun falls directly on cell, electrons are liberated. If both positive and negative sides are attached to electrical conductors, then a closed circuit is formed, by which electrons can be captured in the form of an electric current—that is electricity. Large number of solar cells, electrically connected to each other are called as a module and when multiple modules are wired together to form a large system, then it is called an array.

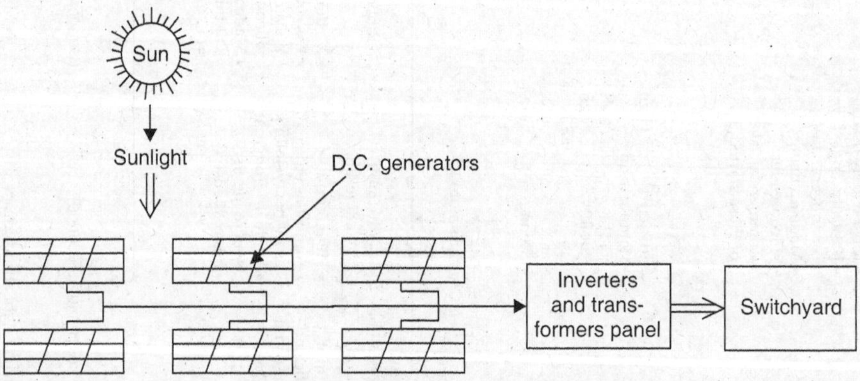

Fig. 4.3: Complete system set-up
[*Source*: Tennesee Valley Authority (TVA)]

However it is the general fact that larger the area of a module or array, large amount of photons from sunlight will be absorbed and more electricity will be produced. They can be used either in series mode or parallel mode as according to the required voltage and current.

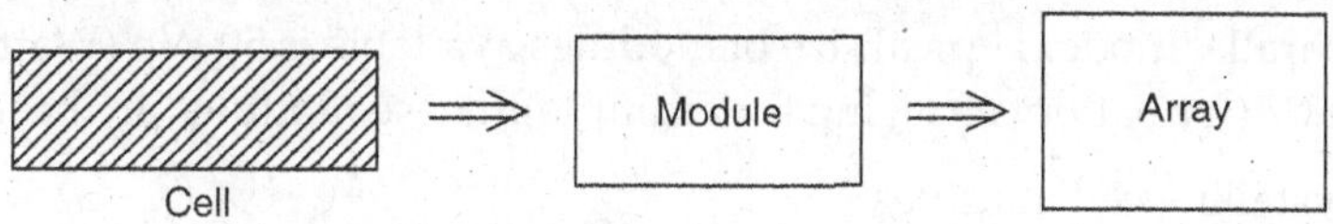

Fig. 4.4: Arrangement of cell, module and array

SOLAR PHOTOVOLTAICS AND DESIGN TRENDS

Introduction

Solar power is an abundant resource available that has the potential for playing a major role in the world's energy supply. Sunlight contains roughly as much energy as in all the fossil fuel. Photovoltaic (PV) is actually the field of study relating to this power. This PV technology is gaining popularity as a mainstream form of electricity generation. Many systems are currently being used at various places of the world, yet this number is insignificant compared to the vast potential that exits for PV as an energy source. These photovoltaics possess lesser cost and higher efficiency, so particularly useful for water pumping systems, health care systems, communication, transport aids and security systems etc.

Although solar energy can be harnessed economically throughout the year, but the angle at which its radiation hits the surface of earth changes during the yearly cycle, so the solar insulation varies or changes. This radiation hits the surface of earth obliquely and solar gain (solar yield) is low. In this case, the orientation of this solar cell panel is very important because the amount of energy captured, depends upon it. The relationship is somewhat complex and only by using sophisticated tracking systems, the maximum energy can be extracted for any given location.

PV Cells, Modules and Arrays

(a) PV cell

A solar cell is basically a photovoltaic device that generates electricity when exposed to light. When light falls on its surface, the electrons get energized in proportion to the intensity of sunlight. When the energy level exceeds a certain point, a potential difference is created automatically across the cell, by which any external load can be supplied. In all these PV devices, silicon is used as the base material. Sometimes other materials such as CdTe (cadmium telluride), CIGS (Copper indium gallium selenide) are also used for improving the efficiency of the device. Typical cell size used is of nearly 15 cms diameter.

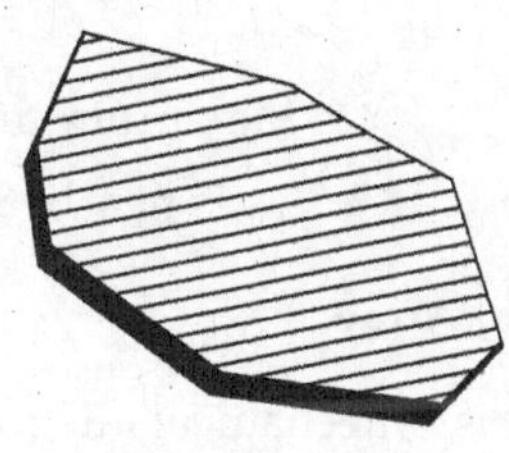

Fig. 4.5: Cell

(b) Module

It is basically smallest non-divisible, self contained and environmentally protected physical structure, which provides single D.C. electrical output. To increase the voltage or current, these are connected in either series or parallel mode. Typical size of modules available is 50 Wp (Watt peak) and creates D.C. electricity at 12V (for battery charging for example). These modules are basically classified as:

(*i*) Monocrystalline cell modules

(*ii*) Multicrystalline cell modules, and

(*iii*) Amorphous, silicon modules

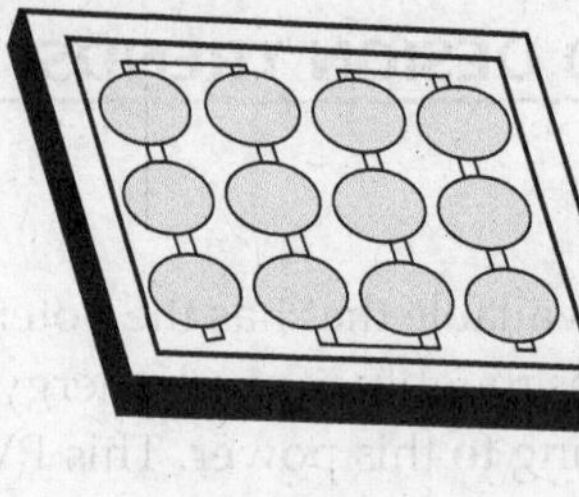

Fig. 4.6: Module

(*i*) **Monocrystalline Cell Modules**

- Maximum cell efficiency is approx. 15%
- Cells are cut from a monocrystalline silicon crystal

(*ii*) **Multicrystalline Cell Modules**

- Maximum cell efficiency is approx. 12%.
- Cells are cut from a Cast ingot of multicrystalline silicon

(*iii*) **Amorphous Silicon Modules**

- Maximum cell efficiency is nearly 6% to 9%
- Made up from thin films of amorphous silicon

(c) Array

It is a mechanical integrated assembly of modules or panels together with support structure such as foundation and other required components to form a free standing field installed unit that produces DC power. It may be classified into three main parts, such as:

(*i*) Flat plate arrays

(*ii*) Tracking arrays

(*iii*) Mobile (portable) arrays

(i) Flat plate arrays

(a) These are kept fixed at a tilted angle (and face towards the equator).

(b) Tilted angle ≈ angle of latitude for site.

(ii) Tracking arrays

(a) These track the path of sun during sunshine period and so harness more sun's power.

(b) Cost of equipments is high and system is somewhat complex.

(iii) Mobile (Portable) arrays

(a) Normally used when equipment being operated is required in different locations, such as some lighting systems or small irrigation pumping systems.

Fig. 4.7: Arrays

All these PV systems are used for stand alone applications. They can be used for battery charging, water pumping systems, etc. Some systems use lenses or mirrors to concentrate direct solar radiation onto smaller areas of solar cells. As the power output is proportional to solar power directed onto the PV cell, so this method is beneficial for reducing area required for collection. These PV systems are also being used for lighting, vaccine refrigeration, electrified livestock fencing and other applications. Solar arrays have been used in power orbiting satellites, other spacecrafts and in remote areas as a source of power for applications such as roadside emergency telephones, remote sensing and cathodic protection of pipelines. The continuous decrement in manufacturing cost (dropping at 3% to 5% a year in recent years) is expanding the range cost-effective uses include road signs, home power generation and even grid-connected electricity generation.

Now let us take a look at comparison of PV technologies, relation between band gaps and efficiencies and world wide PV market respectively.

(a) **Comparison of PV technology:** The main technologies used (available) are single and multi-crystalline Si, a-Si, $CuInSe_2$, CdTe, etc. So the comparison of above technologies can be shown in Fig. 4.8.

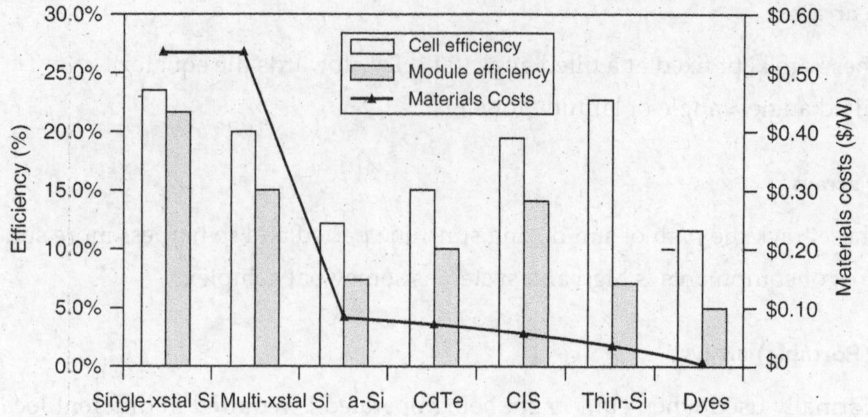

Fig. 4.8: Efficiency and materials cost of different types of cells

(*b*) **Relation between band gap and efficiencies of cells:** Various cells such as Si, cdTe, Si (Amorphous) possesses different band gaps as well as different efficiencies and the relation between the two can be shown in Fig. 4.9.

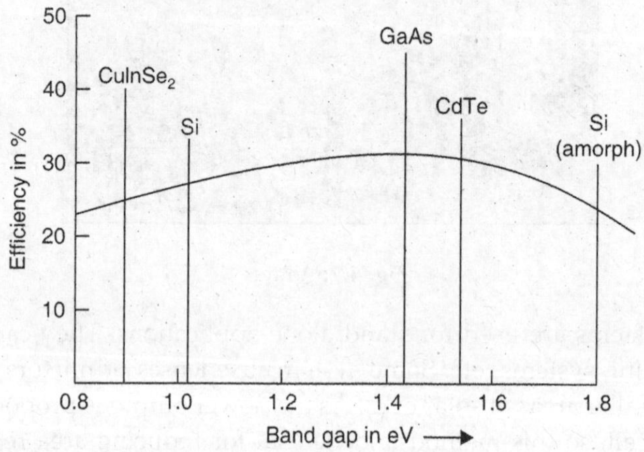

Fig. 4.9: Relation between efficiency and band gap of various cells

(*c*) **World wide PV market:** Solar photovoltaics are by now the fastest growing energy source. The total shipment of PV modules reached nearly 390 peak MW (MW_p) in year 2001, with an increase in the averaged annual growth rate of nearly 10% to 33% over last decade.

Design Trends

In solar photovoltaic system, necessary data as well as informations can be collected as according to the size of system. The main aim of designing is to produce a cost effective and reliable system for a particular application. Since in photovoltaic systems, a number of components are used, so a great attention is also required for the selection of these components. For proper designing a PV system, following procedures are used:

(*a*) First of all total amount of energy is calculated by adding total power needed for all the applications. Also the rated wattage, voltage, operating hours per day (sufficient data) are also arranged as per Table 4.1.

Table 4.1: Representation Summary for Load, AC/ DC Voltage and Energy

S.No.	*Load (AC or DC) Connected*	*AC or DC Voltage Needed*	*Total Hours Used Daily*	*Rated Wattage or Ampere*	*Total Watt Hours per Day*
1.					
2.					
3.					
4.					
5.					
6.					

Thus subtotal AC = (WH/day) DC = – (WH/day)

Also we can adjust AC loads inverter losses $= \dfrac{\text{AC load}}{ndc - ac}$ (WH/day)

(*b*) Total amount of per day load = DC load/day + AC load/day

(*c*) After it, system characteristics for a PV system can be obtained. These system characteristics will be linked with $\eta_{battery}$, $\eta_{regulator}$ etc., for a number of hours during the peak sunlight period.

(*d*) Now, we know that PV array size (peak watt or Wp)

$$= \frac{\text{Total amount of daily load (Wh/day)}}{\text{Peak sunlight period} \times \eta_{battery} \times \eta_{regulator}}$$

Also, Battery capacity

$$= \frac{\text{Daily load (Wh/day)} \times \text{Storage period in days}}{\text{Battery voltage} \times \text{bat.} \times \text{DOD}}$$

where DOD = Battery maximum depth of discharge (from nearly 20% – 80%)

> **Special Points**
>
> (a) $\eta_{battery}$ = Battery efficiency ≈ 75 to 90%
>
> (b) $\eta_{regulator}$ = Regulator efficiency ≈ 85%.

However to achieve the required amount of current and voltage, these PV. Modules can be connected in parallel and series respectively. For a single cell, the operating voltage is nearly 0.45 Vand the current is 270 amp/square cm, so for a 40 sq cm, the value of current would be 1.08 amp and the electrical power output will be 0.49 watts in case of full shiny period of the day. For our problem, suppose, we are designing a solar panel to generate 4 H.P.D.C. Shunt motor, having a rating of 220 Volts (terminal voltage) and 17 Amperes armature current. Since we know that a single photovoltaic module having 30 cells, generates a voltage of 12 V and 1.2 amps current so we have to arrange these modules in series and parallel modes, so that voltage generated must be 220 V and current generated must be 17 A. One fact here is that the voltage of the solar PV module is added in series and adding the PV module in parallel can-increase the current. Thus 20 PV modules are connected in series to generate 220 V and 15 PV Modules are connected in parallel to generate 17 amp.

SOLAR CHIMNEY

A solar chimney can be used as a source of power generation using auxiliaries. This solar chimney with a suitable roof area can produce power nearly 100–200 MW. It actually consists of three main parts, namely glass roof collector, chimney and wind turbines. Thus in other words, it can be said that these auxiliaries are combined in a new manner to have a solar thermal chimney.

In this, by using solar radiations, air is heated under a shallow circular glass roof (open at the periphery) and then the natural ground below it form a hot air collector. There exists a vertical chimney in the middle portion of roof. A large area for air inlet is provided at its base and the joint between the roof and chimney base is made airtight. Since it is a basic phenomenon that suction from the chimney draws in more hot air from the collector and cold air comes in from outside, so by this way a constant up draught is caused by the solar radiations. The energy, which it contains, is converted into mechanical energy by using wind turbines, which are placed at the base of chimney and finally by using conventional generators, it is converted into electrical energy. This solar chimney consists of various benefits. These benefits include:

(*i*) These chimneys produce a large amount of electricity at night also, because the ground under the glass roof absorbs energy during sunshine period and releases it in the night hours.

(*ii*) Very little maintenance cost, due to small number of moving parts.

(*iii*) Reliable and strong compared to other solar generating plants.

(*iv*) Do not need cooling water, thus saving of water becomes a great advantage of solar chimney compared to other sources.

(*v*) Do not necessarily require direct sunlight, but can also accept and utilize diffused radiation under clouded sky.

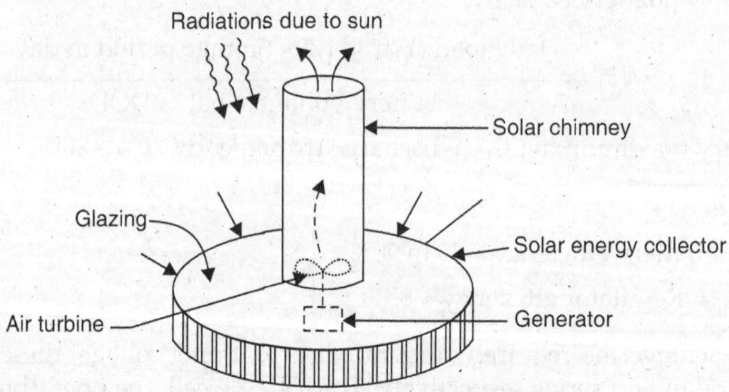

Fig. 4.10: Solar Chimney Power Plant

[*Source*: SUHAS P SUKHATME, Handbook of Solar Energy]

From the economy point of view, large solar chimneys, having collector roof's diameter in kilometers and a height of 1000 meters are needed to be designed compared to smaller chimneys. It can produce power in the range of megawatts at a medium investment level. Actually the demand of future is that a plant should be built now and run for years for demonstration purpose. If it happens, then it will really give sufficient power and attract various countries towards this objective, *i.e.* power generation from solar chimney.

TECHNOLOGY TIPS

All over the world, researchers are continuously trying to produce most efficient solar cells, so in this section, let's take a look at new researches regarding these solar cells.

(*a*) **Pentacene based solar power cells:** In USA, researchers at "Georgia Institute of Technology" have developed a light weight organic solar cell having high efficiency. Actually this "**pentacene**" material is a crystal and this crystal structure of atoms joined together with proper pattern makes it easier, for electricity to move through it, than some other organic material, which are more amorphous. Thus this solar cell is a new invention, giving greater efficiencies.

(*b*) **Plastic solar power cells:** In University of Toronto, researchers have developed a plastic solar cell, which can convert solar power into electrical energy even on a cloudy day. This plastic material uses a new technology called as "**nano technology**" and is a first kind of plastic composite, that is able to harness the infrared portion. Some specially designed nano particles (*i.e.* quantum dots) can be mixed with polymer to make the plastic that can detect energy in the infrared portion of sun's spectrum. Thus by using visible spectrum as well as infrared portion of spectrum, large amount of sun's power can be harnessed, which may provide five times more efficiency compared to current solar cell technology.

SOLAR ENERGY PROGRAMMES BY MNES (MINISTRY OF NON-CONVENTIONAL ENERGY SOURCES)

Solar energy is one of the promising options to make more extensive use of renewable sources derived from the sun. This energy can be used both directly and indirectly. Renewable sources like wind, biomass, hydro power etc. are indirect forms of solar energy. Our country India receives nearly 5000 trillion KWH solar energy per year and the daily average energy incident is nearly 4 – 7 KWH/m^2 depending upon the location. India's rank is first in conducting the solar energy programmes in the world. Various countrywide programmes are being implemented by Ministry of Non-conventional energy sources. These programmes include solar thermal programme and solar photovoltaics programme. Now let us see the brief details of these programmes.

(*a*) **Solar Thermal Programme:** Main purpose is the development of market and commercialization of solar thermal systems so as to meet the domestic as well as industrial requirements. This programme includes the promotion of solar thermal technologies using soft loan programmes, training and education to people awareness creation and help for construction of solar buildings. This program also extends "**Research and Development**" programme for improvement of best emerging technologies. The special output of these programmes are:

(*i*) 10, 00,000 sq. m, solar thermal collector area has been set up.

(*ii*) Large no. of solar steam cooking systems have been also installed.

(*b*) **Solar Photovoltaic Programme:** This programme gives a special option for harnessing sun's energy during shiny periods of the day. The main purpose is to create reliable and cost effective PV systems for rural and remote areas. This program includes set up of solar street lightning systems, PV water pumping systems, PV power plants and other installations using agricultural and related machineries. Under this PV program nearly 85 MW capacity systems have been set up for various uses in our country.

SOLAR GENERATION REPORT (GLOBAL)

Nowadays the global solar electricity market is booming and the cumulative installed PV systems reached at a very high value. The solar generation report highlight that solar energy has the potential to make a major contribution to future global electricity supply. The scenario for year 2020 to 2040 is dependent upon some factors such as market development, market policies etc.

Actually solar generation scenario in India is based on an average growth rate of 35% between 2000 and 2020. It is also stated in the report that the India's PV market would increase beyond 2250 lakh rupees a year by 2020. Thus it will not only provide Indian market growth, but will also add opportunities for jobs for a number of people. The report-projections are as given below:

(*a*) Since solar power is an environmentally friendly renewable energy source, so carbon dioxide emissions will be substantially reduced. It is estimated that these CO_2 emissions would be reduced by 169 million tonnes by 2020.

(*b*) Solar industry will provide jobs to nearly 2.25 million people by the year 2020 in installation, maintenance and marking areas and will be proved as a source of happiness for poor people.

(*c*) By (Targeted year 2020), photovoltaic systems would produce approximately 282 Terawatt hours of electricity in the world, as according to this report. Also by 2040, the solar power's contribution would raise to 21% of world's electricity output production. It would definitely place solar power firmly on the map as an established source.

MAKING SOLAR ENERGY MORE VIABLE

By using a system, developed at the U.S. "Department of **Energy's Oak Ridge National Laboratory**" (ORNL), solar energy could get a mega boost, effectively gaining improvement over conventional technology. The "full-spectrum solar energy system" harnesses clean and abundant energy from sun and utilizes that energy in a much more efficient manner than is possible with traditional solar systems. Actually this technology departs from conventional approaches, that focus on using solar energy for a single purpose so as to generate power.

Actually the amount of electricity consumed can be decreased by using the visible portion of light spectrum and the other portions of the spectrum can be used to generate electrical power. The special system was developed by Muhs and industry partners, which uses novel roof mounted two axis tracking concentrator that separate the visible and infrared portions of sun's rays. However when optical fibres are used with large diameter, then it distributes visible light to interiors of buildings and thus interior infrared portion of sun's spectrum is converted into electricity. Thus, the total approach provides advantages of concentrating photovoltaic cells in the near infrared portion of the spectrum, where they are generally more efficient.

Visible light portion must be mixed with artificial light to maintain uniform lightning-even in the failure of sunlight. The researchers actually dubbed this combination of natural and artificial light, called as "**hybrid lightning.**" It gives the advantage of better use of direct sunlight and the reduction in the consumption of energy during peak demand. This hybrid lightning is making tough competition with conventional day lightning and solar technologies. Integrated with full spectrum solar energy systems, it provides a number of advantages. Thus it is clear now that the potential for commercial application of a full spectrum solar energy system is significant, so the overall analysis indicates that by using this hybrid approach to solar energy there is an ample opportunity to achieve simple paybacks of less than five years.

ROLE OF NANOTECHNOLOGY TO INCREASE THE EFFICIENCY OF SOLAR CELLS

As we all know that the production costs of electricity using conventional sources is continuously increasing day by day but solar cell's cost is continuously decreasing. In future we can say that solar photovoltaics will be proved as a cheaper source of electricity production.

It must be clear that the cost of electricity production of solar cells will decrease, but however if we can improve the solar cell efficiency by any method then it will be proved as a boon for solar cell market. Thus at this stage, we must discuss some important ways so as to generate efficient solar cells. These methods include:

(*a*) Multijunction devices; (*b*) Important techniques using TiO_2, DSSc and nanotubes; (*c*) black silicon surfaces; (*d*) Photons with more than twice the band gap energy; (*e*) Titania nanotubes; (*f*) hybrid nanomaterials; (*g*) Using the full spectrum of sunlight including infrared portion of spectrum; (*h*) Semi conductor quantum dots; (*i*)Using nanoparticles in Octillion "Nanopower Window".

All these methods will be discussed here one by one.

(*a*) **Multijunction Devices:** It is also called as cascade or tandem cell. A multijunction cell can be made up by using two different ways, namely the mechanical stack approach and other monolithic approach. In the case of mechanical stack approach, two individual solar cells are made independently. One of them is made up using high band gap and the other is made with a lower band gap. In this two cells are mechanically stacked, one on the top of other. Likewise in second approach, first a complete solar cell is prepared and then the layers for the second cell are grown or deposited directly on the first.

All these individual cells are arranged in such a way that sunlight falls on the material, which has the largest band gap. Now the photons, which are not absorbed in the first cell, are transmitted to second cell, which absorbs the higher energy portion of remaining solar radiation and remaining transparent to lower energy photons. This absorption process continues through to final cell, which has the smallest band gap. These cells possess efficiencies of more than 35% under concentrated sunlight.

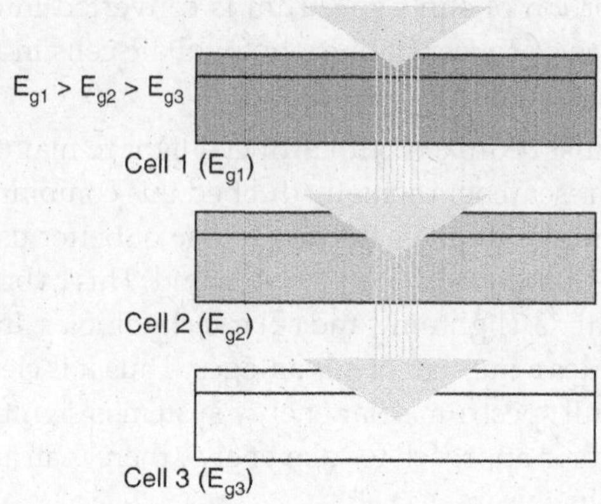

Fig. 4.11: Multijunction device, stack of Individual single junction cell
[*Source*: Solar Energy technologies programmer, U.S. Department of energy]

(*b*) **Important techniques using TiO_2, DSSc and Nanotube:** These three technologies can be summarized as follows:

(*i*) In the first way, we put the titanium dioxide in as a layer between two parts of a cell. The first part of the cell, where the electrons are liberated and the second part where they are collected for dispatch into wider world. Now by keeping the resulting sandwich in the oven for a few minutes at 150°C, plastic layer becomes crystalline. Since it is easier for electrons to move through crystalline structure, it can easily improve the efficiency of the whole process.

(*ii*) In the second case, DSScs (Dye sensitized solar cells) are exciting variant of the most efficient and stable of the excitonic photovoltaic devices. Actually DSSCs is a thick titanium dioxide (TiO_2) nanoparticle film, which gives a large surface area for absorption of light-harvesting molecules. In this electrons are passed via. a semiconductor to an electrode, through which they leave the cell. When the TiO_2 surface is modified with dye molecules, then it absorbs both visible and infrared wavelengths. In this case, scientists are able to increase the efficiency above (1%).

(*iii*) In the third case, researchers are able to improve the efficiency of cells using nanowires. In this we cover the surface of a cell (made up of cadmium sulphide, zinc oxide and TiO_2) with nanotubes, so that these tubes stuck up from the surface like hairs. Now the tubes ease the passage of liberated electrons from the cell to the electrode that collect them. This technique doubled the efficiency of cell at ultraviolet wavelengths.

(*c*) **Black Silicon Surface:** Another important technique to make solar cells more efficient is to use silicon surfaces rendered black by pits. In reality, flat silicon surfaces are highly reflective, so a greater portion of light is reflected and lesser amount of electricity is produced. In order to have greater efficiencies of these solar cells, antireflective coatings must be used, which actually reduce the amount of average reflection in the wavelengths of light.

Now a days, scientists have developed a technique, which creates the textures using wet chemical processes. Actually in this procedure grains of gold, only nanometers large, are deposited onto a flat silicon surface. Now with a solution of hydrogen peroxide and hydrofluoric acid, the silicon between the areas covered by the clusters is etched away. The gold nanoparticles show catalytic action "behaving like drills into the surface". On the surface, 50-100 nanometer high pits are formed by the areas covered by the gold, while the silicon between the gold makes up the bumps. Now with a solution of iodine and potassium iodide, the nanoparticles are removed. Over all we can say that it is being proved as a very easy way to achieve an extremely non-reflective silicon surface and an important breakthrough on an important issue.

(*d*) **Photons with more than twice the band gap energy:** As we know that solar cell works by converting a photon of sunlight into an electron-hole pair, which generates electricity. In relation to this another technique to increase solar cell efficiency is to get photons with more than twice the band gap energy to produce two electron hole pairs. It is actually achieved by splitting photons, which converts them to lower energies.

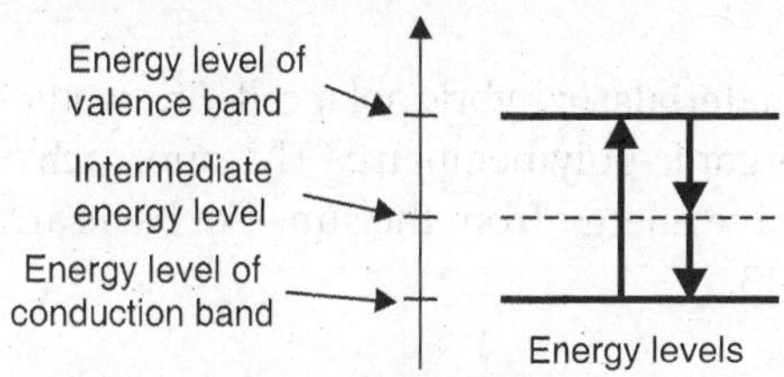

Fig. 4.12: Energy levels

By using proper down-conversion materials, solar cell efficiency can be improved and the best efficiency can be obtained when the down converter material is kept on the front or rear surface of solar cell, in front of a mirror that bounces radiation back. It minimizes the amount of energy lost by absorption in the down-converter. In addition to this, by using a material having high refractive index for the down-converter, the efficiency of cell can be greatly improved, because it enables more emitted photons to be directed into the solar cell.

(*e*) **Using Titania Nanotubes:** A new way to harness the sun's power is by using highly ordered arrays of titania nanotubes for hydrogen production and increased solar cell efficiency. As we all know that an important prospect for an efficient renewable resource is solar energy either to generate hydrogen or power solar cells. By using highly-ordered titanium nanotube arrays, a lot of hydrogen per photon, out of the system can be obtained. If its band gap could be shifted successfully into the visible spectrum, we would have commercially practical means of generating hydrogen by solar energy. Actually these nanotube arrays provide excellent pathways for electron percolation, in effect acting as electron highways for directing the photo generated electrons to where they can do useful work.

A solar cell made of titania nanotubes and natural dye may be proved as an answer to make cost effective solar electricity production. Using an easier approach, the process starts with a glass piece that has been coated with fluorine-doped tin oxide, which is now followed by a 500 nanometers thick titanium layer. Now layer is placed in an acidic bath undergoing a mild electric current. It causes parallel nanotubes of titanium dioxide to form and grow to a length of about 360 nm. Then tubes are heated in an oxygen-rich environment to crystallize them. Thus, the process transforms the opaque titanium coating into a highly ordered array of transparent titanium dioxide nanotubes. Finally, the nanotube array is coated with a commercially available photosensitive dye, forming the negative electrode of the cell.

Thus, the dye-coated nanotubes make up the negative electrode and a positive electrode seals the cell, which contains an iodized electrolyte. When sun rays passes through glass, the energy falls on the dye molecules and electrons are feed producing current. Hence an important thing is that by using thicker initial coatings, more electrons or more electricity would be produced by longer tubes, because the thickness of titanium layer constrains the height of the nanotubes. Thus by increasing the length of transparent nanotube arrays, we can get more conversion rate with these cells, using a relatively easy fabrication system, that is commercially viable, so we can hope that by using nanotubes, efficient solar cells could be produced.

(*f*) **Using hybrid nanomaterials:** In hybrid solar cells, inorganic-semiconductor nanomaterials get dispersed in an organic-polymer matrix. This approach provides light weight, efficient, robust and inexpensive energy from the sun. The basic arrangement of hybrid solar cell is as given in Fig. 4.13.

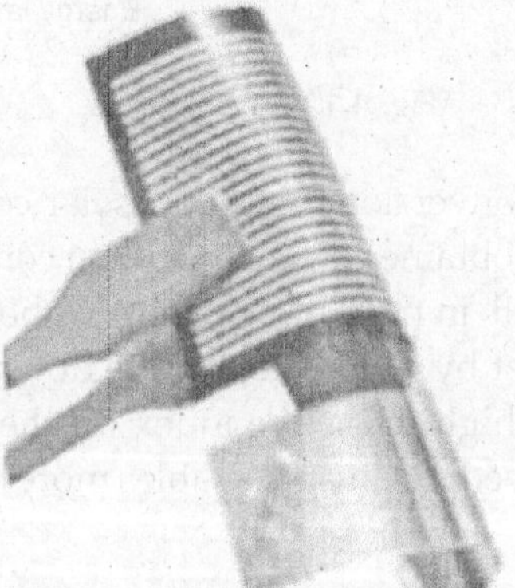

Fig. 4.13: Hybrid solar cell

Efficiency of polymer based devices is poor and fortunately new nanomaterials have been producing remarkable improvements in these devices. The best results have been produced by nanoscale crystalline semiconductors such as cadmium selenide and nanostructured forms of carbon. The performance of hybrid solar cell have been improved by tuning the properties of the nanomaterial for absorbing sunlight. When material dimensions become comparable to spatial extent of the electrons that occupy it, the materials start to exhibit quantum-confinement effects. As the size of particle is reduced, light being emitted (Character of the semi conducting gap, or the wavelengths at which the crystal can absorb light) shifts to higher energy. The nanomaterials used in hybrid approaches exhibit optical absorption below the conducting polymer band gap and so allow these composite devices to absorb a much larger portion of solar spectrum. Thus in brief we can say that polymer solar cells provide, a nanomaterials approach for improving efficiency of a device.

(*g*) **Using Full Spectrum of Sunlight:** In case of solar cells, to achieve high power output, solar devices must take advantage of much of the solar spectrum from infrared to visible light to UV, as the photons absorbed by a solar cell directly impacts the power output. In reality, solar radiation covers wavelength as short as 300 nanometers and as long as 4045 nm. The amount of incoming photons across the UV, visible and IR spectrum are nearly 3%, 45% and 52% respectively.

An important case here is of PCE (Power Conversion Efficiency). Ability of material to efficiently absorb solar light across a broad layer of wavelengths impacts the PCE potential of solar cell.

Silicon devices absorb and convert solar energy up to about 1050 nm covering approximately 75% of the total photon flux from sun.

- **Newer materials and special architecture:** Now a days some researchers at Princeton and USC have found a new record in solar cell, that is responsive to light in the near-infrared range of solar spectrum. Actually an architecture is used which includes C 60 (Buckey balls), copper (Cu) and tin phthalocyanine (Cupc and Snpc). For shorter wavelengths (450–550 nm), C 60 is a good absorber, beyond UV light. Similarly Cupc is well suited from 560 nm to 750 nm and snpc is a strong absorber for wavelength 628 nm to 1000 nm. In a good performance organic PV cell, the active organic layers (Cupc and C 60) are sandwitched between a transparent conducting oxide layer (TCo) on one side and a blocking layer and metal electrode on the other. It was hypothesized that a very thin layer could be used so as to absorb NIR light without hindering mobility. Princeton researchers would have substituted the Snpc for the Cupc in this structure to improve long wavelength absorption as Snpc is a stronger absorber compared to Cupc over nearly its entire range. As the Snpc layer was thinned, however it was found that electrical shorts began to form between the C60 material and the Tco layer, as the Snpc layer was no longer continuous. Now the successful solution was created by using Cupc as a "wetting" layer prior to deposition of very thin, Snpc layer. Its strong absorption characteristics allowed the device to efficiently perform into the near-infrared portion of the solar spectrum, despite the thinness of Snpc.

  Thus, the maximum external quantum efficiency of 35% was achieved by the device, so this latest device proves that from the IR and near IR portion of the solar spectrum, greater power can be achieved. In fact, this approach has the potential to double the power output of organic solar devices with power harvested from the near IR and IR portions of sun's spectrum, greater power can be achieved.

- **Indium gallium nitride: An important material to harness sun's power.** In the solar cells, the most efficient semiconductors are alloys made from aluminium, gallium and indium, with elements like nitrogen, arsenic etc. and the best semiconductor material is Indium gallium nitride, which may absorb more of sun's energy than today's best alternatives. By varying the indium/gallium ratio in different layers, scientists were able to tune InGaN's band gap to match the criteria exactly. Now because the band gap range matches the solar spectrum so perfectly, a cell could use multiple layers tuned to band gap across the range to squeeze out even more energy.

Recent research report shows that a team led by Wladek Waly kiewicz, working with researchers at Cornell University and Ritsumeikan University, Japan have found that single system of alloys having indium, gallium and nitrogen can convert, the entire sunlight spectrum from the near infrared to far ultra violet to electricity. With a band gap of 3.4 ev, invisible UV light is emitted by gallium nitride, but when some of the gallium is exchanged for indium, various colours like violet, blue and green are created. In fact longer wavelengths could be produced by the same alloy, if the proportion of indium was increased, soon it was established that as the proportions shift from indium toward gallium, band gap width of alloy increases in a smooth and continuous manner, until having covered every part of the solar spectrum. The use of indium gallium nitride looks promising for solar cells, because this material system has an optimal gap range for the conversion of almost the entire solar spectrum and is very tolerant with regard to lattice mismatches. A team of Wladek Walukiewicz and his colleagues performed absorption tests on extremely high quality InN and $In_x Ga_{1-x}$ N films grown under carefully controlled conditions. It was found that pure InN's direct band gap is 0.7 eV and mixing InN with GaN to form $In_xGa_{1-x}N$ can produce materials, whose band gap can be continuously varied from 0.7 ev-3.4 ev. Now for the range 0.7 ev –3.4 ev, we have following wavelength of solar spectrum using the formula $\lambda = hc/E$.

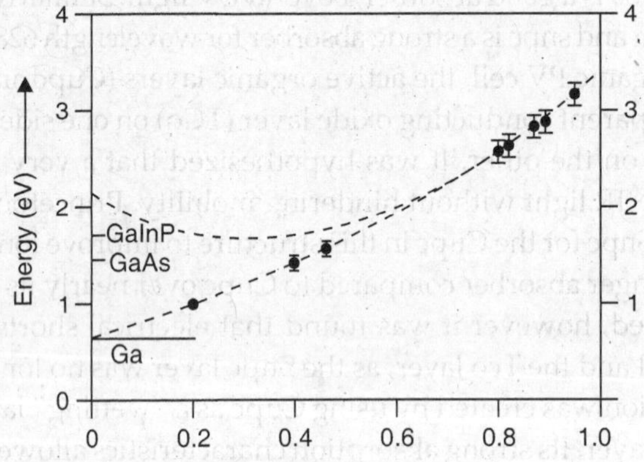

Fig. 4.14: Ga Content in In_{1-x} Ga_x N alloy and perfect matching of band gap of In_{1-x} Ga_x N to solar spectrum

[*Source*: 1. (http: // www.lbl. gov/science articles).

2. J. Wu, W. Walukiewicz and colleagues, "Unusual Properties of the fundamental band gap of InN" Appl. Phys. Lett. 80, 3967- 3969 (2002)].

Table 4.2: Relation Between Wavelength and Band Gap

Band gap	0.7 eV	1 eV	2 eV	3 eV	3.4 eV
Wavelength	1615.7 nm	1137 nm	565.5 nm	377 nm	332.6 nm

Similarly $In_x Al_{1-x}$ N system has band gap range from 0.7 – 6.2 eV; so these related materials may be used for both solar energy conversion and in other applications in the near-IR to far UV region of spectrum. Thus this semiconductor system has an almost perfect match to entire solar spectrum, by which more electricity can be produced. Overall we can say that these materials and semiconductor systems will be useful for achieving high solar cell efficiencies.

(*h*) **Semiconductor quantum dots:** Quantum dots or nanoparticles are basically semiconducting crystals of nanometer (a billionth of a meter) dimensions. They possess quantum optical properties, which are absent in the bulk material and can easily be molded into a variety of forms. In reality, another starting point for increase of conversion efficiency of solar cells is the use of semiconductor quantum dots. The band gaps can be adjusted to convert longer wave light and so efficiency of solar cells can be increased.

The Carnot-limit on the conversion of sunlight to electricity is nearly 95% as opposed to the theoretical upper limit of 33% for a standard cell. The best scenario would provide a low cost semiconductor material that could have its band gap tuned for optical performance allowing manufacturer to control the absorptive properties of the solar cell and the line of quantum dots have been designed for exactly this type of tunability.

These quantum dots are designed to absorb preferentially a part of the spectrum. The synthesis of quantum dots of Cdse and $CuInSe_2$ for use in intermediate band gap solar cells are being investigated by NASA Glenn research centre. Using quantum dots in a solar cell to create an intermediate band will allow the harvesting of a much larger portion of available solar spectrum. Therefore quantum dots give a much more exact method of matching the band gap of solar cell material to the optimum band gap for energy conversion, giving greater efficiencies.

(*i*) **Using Nanoparticles in Octillion's 'nanopower window':** A new research related to solar cells, shows that when silicon nanoparticles are placed on a silicon solar cell, then the power output can be boosted and energy wasting heat can be substantially reduced. The experiments show that when silicon nanoparticles are kept on the top of solar cells, then large voltage enhancement with increase in power ($\approx$ 60 to 70%) in the ultraviolet–blue (UV) range is obtained.

According to Mr. Harmel S. Rayat (President and CEO of Octillion corp.) the exceptional power performance of these silicon nanoparticles is a substantial achievement and is especially significant. This technology of nanoparticles is the key to the development of Octillion's transparent glass windows, capable of producing electrical power. In addition to this, these nanoparticles are also able to convert same UV components (that typically cause damage and create wasteful heat) into useful electrical output. Thus in brief it can be said that this process not only the power is increased, but also UV damage to solar cells is decreased and finally lifespan of solar cells is increased. Thus, when thin films of silicon-nanoparticles are sprayed onto silicon substrates, UV light is absorbed and converted into electrical current. With proper circuit arrangements, the film acts as nano-silicon photovoltaic solar cells that convert solar radiation to electrical energy.

Thus in conclusion, it can be said that in future, if the power of sun is harvested with improved techniques as described above, then definitely a great amount of electricity would be obtained or in other words, we can say that in future due to the production of highly efficient solar-cells, solar cell technology will be proved as one of the better sources.

In reality, wind power is called as a renewable energy source and in India, this power is continuously providing electricity to a number of poor homes in rural as well as in urban areas. So as to harness power of wind, wind farms are made in coastal areas, at the top portions of rounded hills, open plains and gaps in mountain places, where wind is strong and reliable. To be worthwhile, average wind speed of nearly 25 km/hour is needed.

In India, nearly 4.5% of total installed generating capacity is of wind power. This power makes the nation's wind energy programme, the fifth largest in the world. One important thing is that the net wind energy potential in India is more than 45 thousand megawatt and the states having high wind power potential are Gujarat, Andhra Pradesh, Kerala, Rajasthan, Madhya Pradesh, Maharashtra, Tamil Nadu etc.

- **Working of wind power arrangement:** Sun provides radiation to heat the atmosphere and the earth. Since our atmosphere is heated unevenly, so some patches become warmer compared to others. These warm patches of air rise, cooler air blows in to replace them and finally we feel a wind blowing. By using a tall tower, propeller and other auxiliaries, a plant can be set up to produce electrical energy. The basic arrangement is as given in Fig. 5.2.

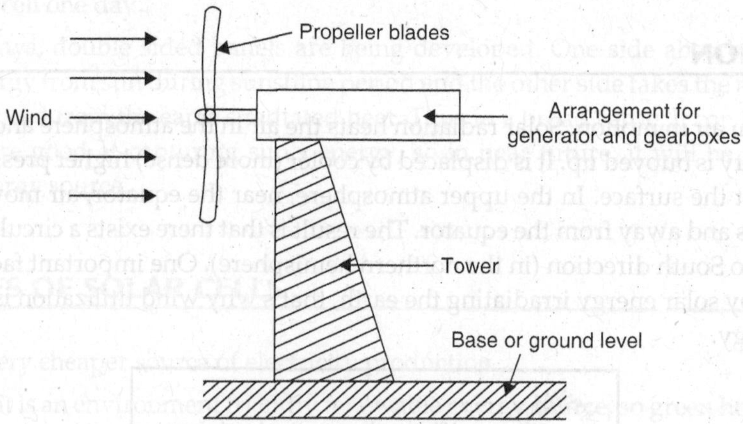

Fig. 5.2: Wind power system

Wind moves in a particular direction and blows the propeller round, which in turn, moves the generator attached and finally electrical energy is produced. However if large amount of electricity is needed, then a number of towers are used to form "**Wind farm**". The simple arrangement of windmill with its components is as given in Fig. 5.3.

The simple concept is that more the towers, more the wind, large the propellers, then more electricity we can generate. In the case of wind power generation, large propellers are used so as to have energy from largest possible volume of air. The blades can be angled to "fine" or "coarse" pitch, to cope with varying wind speed and the generator and propeller arrangement can turn to face the wind.

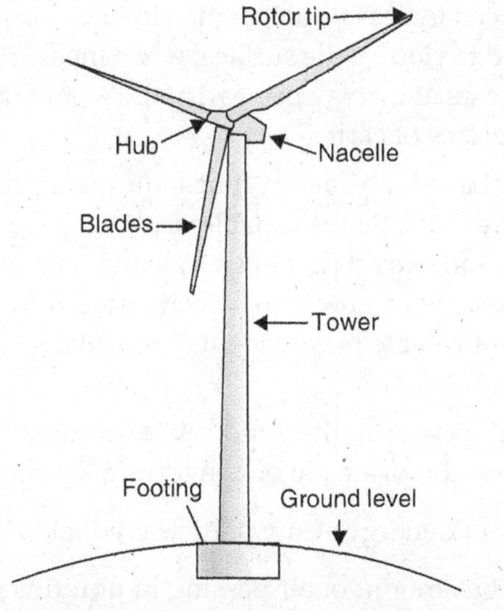

Fig. 5.3: Windmill and its components

However the summary of all these things mentioned above can be represented by Fig. 5.4.

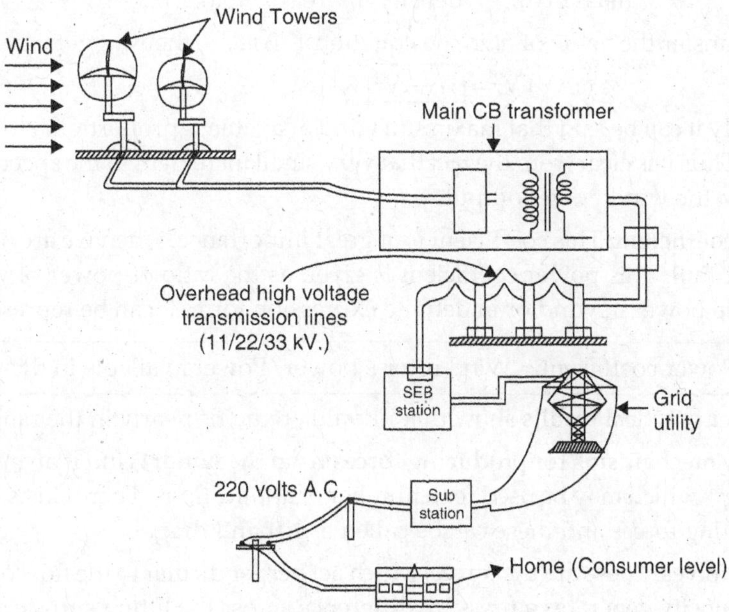

Fig. 5.4: Arrangement of supply of electricity from windmill to home

SOME ASPECTS REGARDING WIND POWER

(*a*) **Wind nature:** As we all know that air circulation in the atmosphere is due to the non-uniform heating of earth's surface by the sun. Air expands and is forced upward by cool, denser air, which flows in from surrounding areas causing a wind. One important nature of wind is that its speed increase with height. However, practice shows that they are nearly 25% more compared to close of the surface, when measured at a height of 10 meters. Also at a height of 60 meters, they may be nearly 30%–60% higher due to the reduction in the drag effect of the surface of earth.

(*b*) **Wind power:** Wind has a lot of energy, but some means must be provided for harnessing them. Actually any device, that is capable of slowing down the mass of moving air (for example, propeller) can extract part of energy and convert it into useful work. Although wind has a great amount of power to be converted into electricity, but a 100% efficient aerogenerator cannot be able to convert 100% available energy, but would only convert maximum of 60%.

By using "**kinetic's**", power in the wind can be obtained. Windmill (used for electricity generation) works on the principle of converting kinetic energy of wind to mechanical energy. Since the kinetic energy of any particle is equal to $\frac{1}{2}$ mass times square of velocity *i.e.* $1/2\ mV^2$. Also the amount of air passing in unit time, through Area *A* with velocity *V* is A.V. Also we know that

$$\underset{\text{mass of air}}{m} = \underset{\text{density}}{e} \times \underset{\text{area}}{A} \times \underset{\text{velocity}}{V} \qquad \text{... (1)}$$

If we transfer the value of *m* in the equation $1/2\ mV^2$, then we get

$$\tfrac{1}{2}(eAV)\,V^2 = \boxed{\tfrac{1}{2}\,eAV^3} \text{ watts.}$$

So finally it can be said that maximum wind available is proportional to the cube of wind speed. Thus it is clear from the fact that very small increment in the speed can have a large effect on the wind power output.

(c) **Power coefficient:** This coefficient is of great importance, when we are dealing with rotor of wind mill. This power coefficient is given as the ratio of power of wind rotor to the available power in wind or in definite expression form, it can be represented as:

Power coefficient = Wind rotor's power/Power available in the wind

However practical results show that its value remains nearly in the range of 0.35 – 0.59.

(*d*) **Primary mechanisms for producing forces from the wind:** Wind is an intermittent source of energy, which may be used for a number of applications. There exists two mechanisms for creating forces and these can be called as **lift and drag.**

(*i*) **Lift forces:** These are the forces, which act perpendicular to the flow of air. Any change in velocity generates a pressure difference across the lifting surface and this pressure difference creates a force, that begins to act on the high pressure side and moves towards the low pressure side of the lifting surface, called an air foil.

(*ii*) **Drag forces:** These are the forces, which act in the direction of flow. For a stable and efficient operation a wind turbine blade needs to function with as little drag as possible. One important aspect is that this drag type of force dissipates energy also.

The design of the system must be in such a way so that it can produce high value of lift/drag ratio. The design of each wind turbine specifies the angle at which the air foil must be set to achieve the maximum lift/drag ratio.

BASIC COMPONENTS OF WIND ENERGY CONVERSION SYSTEMS

For converting wind energy into electrical energy, turbine-generator set is required. The output of this generator is further connected to the load side or power grid as according to the application requirement. Now let us discuss some basic components of the above system.

(*a*) **Rotors:** In the systematic arrangement, two types of rotors are preferred, namely horizontal axis rotor and vertical axis rotor. Both these rotors are used for different types of machines, as according to requirement for a particular application, but generally vertical axis machines are preferred, because they can be operated in all wind directions.

(*b*) **Windmill head:** It is one of the most important part of windmill arrangement. The purpose of windmill head is to

(*i*) Place the bearings of rotor properly.

(*ii*) Support the rotor.

(*iii*) Provide accommodation for all control mechanisms associated with system.

(*c*) **Towers:** The tower is the basis for overall system. Various types of these towers are used for different applications. They include:

(*i*) Shell-tube type tower

(*ii*) Truss tower

(*iii*) Pole tower

(*iv*) Reinforced concrete.

Although all these types of towers are used for various applications, but Truss tower is mostly preferred due to its low cost, potential stiffness and easy availability.

(*d*) **Controlling arrangement:** Controlling arrangement used must be reliable in nature and it must perform following functions as given below.

It must continuously monitor the status, data, storage and data computations etc and take care of auxiliary equipments connected in the system. In addition to this important problem such as shutdown and cut out due to malfunction or very high wind velocity, must be completely eliminated.

For the purpose of controlling, many components, such as sensors (may be electrical or mechanical), relays, various analog circuits with microprocessors, logic modules, actuators (may be pneumatic or electrical), transducers and electronic equipments must be used in association with necessary setups.

(*e*) **Yaw control:** Wind turbines used in the conversion system must be yaw active *i.e.* with the change in wind direction, a motor rotates the turbine slowly about the vertical (yaw) axis so as to face the blades into the wind. The main function of this controller is to sense the speed of wind, direction of flowing wind, shaft speeds and torques at various points. It must sense and control the power output and generator temperature.

(*f*) **Turbine and generator set:** Turbine is the part, used as a prime mover for turbine-generator set. In the case of wind turbines, axis may be vertical or horizontal, its direction may be either upwind or downwind and blades may be either 2 or 3 in nature. Thus there are

rate continues to remain at a constant level. So let us take a tariff schedule for wind energy project (for a number of years).

Table 5.2: Tariff Schedule for Wind Energy Project

S.No.	*Financial year of CoD*	*Rate in rupees per KWh*
1	2010–11	3.21
2	2011–12	3.26
3	2012–13	3.31
4	2013–14	3.36
5	2014–15	3.36
6	2015–16	3.36
7	2016–17	3.36
8	2017–18	3.36
9	2018–19	3.36
10	In subsequent years	3.36

Likewise other agencies, such as Indian Renewable Energy Development Agency and financial institutes are playing a major role in India for promoting renewable energy projects (for wind power). Thus, it can be hoped that in near future, India may be at the top position of the world in power production from wind.

Despite the location in the tropics, India is endowed with substantial wind energy potential due to its geographical location, within the area covered by the great Asian monsoon circulation system. The monsoon wind remains in great strength during June to September for providing great energy potential in India. A recent report shows that those areas having wind speeds greater than 0.18 km per hour (minimum requirement is 0.18 kmph) are in some parts or states of country (or discussed in previous section). In addition to this, we must take a look at installed wind capacities in various regions and countries, which is given in Table 5.3.

Table 5.3: Global Installed Wind Power Capacity (MW) [Up to end 2010]

S. No.	*Region/Country*	*Installed wind capacity*
1	Africa & Middle East	1079
2	Asia	61087
3	Europe	86279
4	North America	44189
5	China	44733
6	Japan	2304
7	Germany	27214
8	France	5660
9	United Kingdom	5204
10	USA	40180

WIND POWER IN FUTURE

According to latest survey, the total amount of electrical energy requirement in India will be more than 9.88 Million KWH by the end of 2014. It has also been estimated that in next 10 years; nearly 8000 Megawatt of wind farm capacity would be installed in some states like Karnataka, Andhra Pradesh, Tamil Nadu etc. In the western region, the energy requirements by March 2014 will be more compared to southern region of India.

The commissioning of wind farms over above mentioned states, during next ten years could generate at least 6000 Megawatt. This if serious efforts are done in this direction, by the cooperation of public and private sectors, then obviously a large amount of generation from wind will provide a boon for Indian market. Now the important issue of discussion is that, what must be the role of locally available people, government and engineers? The local people must be made aware about the potential of wind energy. They must be explained about the advantages of such power, so that they may help in the establishment of such projects. Thus those people may provide a great help for wind power industry in India.

Similarly government must provide concessions like grant of loans, tax concessions etc. The state government must run education programmes for the farmers and agriculturists. They must be convinced by the government, that setting up of hundreds of wind generators will not be harmful to agricultural production as well as for other activities of rural areas. Similarly on the engineering side, to promote indigenous manufacture of wind-electric generators and their accessories, their testing, certification, repair and maintenance of a number of wind farm complexes, all possible steps must be taken, by the engineers.

TYPES OF WIND MACHINES

Wind machines form the basis of harnessing electrical power. These are mainly divided into two types namely horizontal axis machine and vertical axis machine. Now let's have a brief discussion regarding these machines.

(*a*) **Horizontal axis machine:** These are the machines, which have their blades like airplane propellers. These machines may be as tall as 20 story building. It consists of 3 blades arrangement, that span 200 feet across. The largest wind machines in the world have blades longer than a football field itself. Since these machines stand tall, so grasps more wind and thus more amount of power is produced (Fig. 5.6).

Nowadays, the engineers are using the bicycle wheel windmill. This wheel has narrow rims and wire spokes. Power from these bicycle wheel rotors can be taken through a belt drive or directly from the rim (Fig. 5.7).

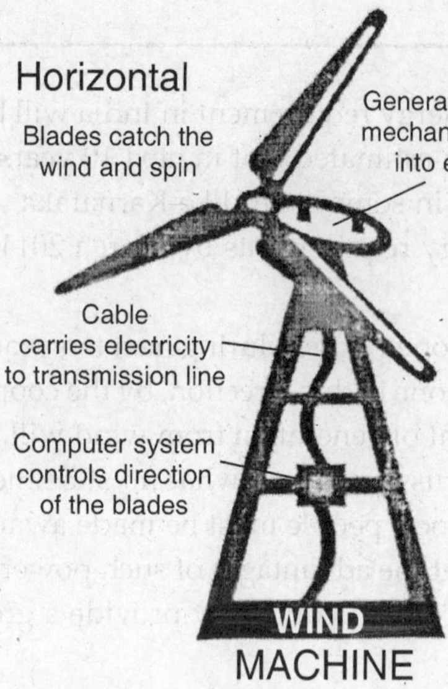

Fig. 5.6: Horizontal wind machine set-up

Fig. 5.7: Horizontal axis machine

(*b*) **Vertical axis machine:** These machines possesses their blades that go from top to bottom and look like giant egg beaters. These machines stand 100 feet tall and approx. 50 feet wide. Nearly 5% of machines, used are vertical axis machines. So as to reduce the cost of turbine and system, the above machine (having rotors that spin or vertical axis) is preferred. One important fact related to this machine is that the machine can easily be installed on the ground level without tower structure and its rotor blades are designed to bear any type of wind from any corner of the compass.

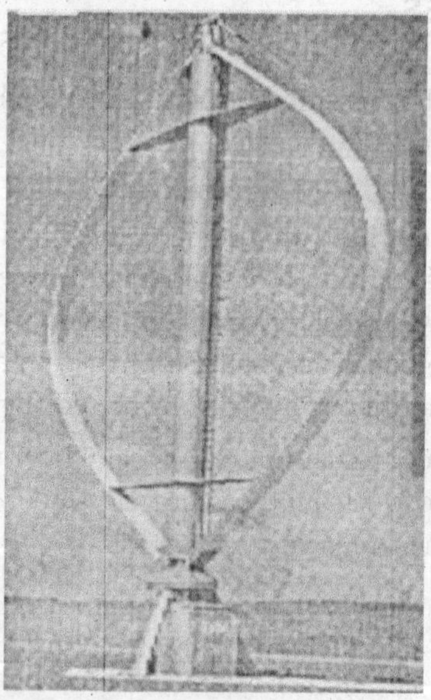

Fig. 5.8: Vertical axis machine

COMPARISON BETWEEN FOSSIL FUELS AND WIND

Nowadays, when wind power has been emerged as a beautiful and reliable source of power generation, each of us must be aware about the advantages of wind compared to fossil fuels. So let us see the detail representing basic comparison between the two, as mentioned below:

(*a*) **Limitation on availability:** Air is an inexhaustible resource, whereas fuel is limited in reserves and expected to get completely exhausted in the coming years.

(*b*) **Use in production:** The former has zero emission but latter produces greenhouse gases resulting in dangerous effects on atmosphere.

(*c*) **Availability:** Air is available and useable everywhere because it exists naturally but fossil fuels are required to be procured and made usable through time consuming, laborious and environmentally damaging processes.

(*d*) **Geo-political implications:** Air actually decreases our reliance on oil and thus provides national security. On the other hand, over-reliance on oil has undermined our energy security [For example → Iraq war of 2003].

(*e*) **Pollution factor:** Air obviously has no bad effect on environment. However if we implement wind energy harnessing technologies, using wind turbine-generator set, then pollution saving takes place in terms of following manner:

(*i*) Sulphur–dioxide (SO_2): → 2 to 3.2 tonnes
(*ii*) Nitrogen oxide → 1.2 to 2.4 tonnes
(*iii*) Carbon dioxide (CO_2) → 300 to 500 tonnes

} For an average output of 4 MWH/day.

On the other hand, fossil-fuels are responsible for environmental degradation.

WIND ENERGY AND PHOTOVOLTAIC (AN INTEGRATED APPROACH)

We know that wind power is an efficient and cost effective form of renewable energy available in certain parts of India. Wind energy is actually environmental friendly at a time, when there is a growing concern about population and climate changes. In reality, one of the target to supply power is to have maximum output, which rarely exceeds the load requirement (2000 watts for example). Now for example, we use an integrated approach. If the rating of wind turbine is 1000 watts, at 20% capacity factor at the site (average nearly 200 watts for example) and also if the solar array system is rated at 1000 watts at 30% capacity factor (average nearly 300 watts), then at the wind speed of 9 meters per second and sun-shine period, 100% power would be delivered from the wind turbine and solar panels.

Likewise for an off grid or stand alone system, other option may be to size the various components to meet out the local load and back up the wind and solar generation using battery storage and diesel generator. Thus, the battery back up and generator will be needed and these both will ensure about the full electricity supply all the time.

Thus in brief it can be said that the main aim of integrated approach is to have a cheaper and secured supply source to meet the electricity needs. However if this approach is utilized for

Chapter 6 *Biomass Energy*

INTRODUCTION

Biomass is basically solar energy stored in organic matter. With the growth of trees and plants, process called as "**photosynthesis**" uses energy from the sun to convert CO_2 into carbohydrates such as sugars, starch and cellulose.

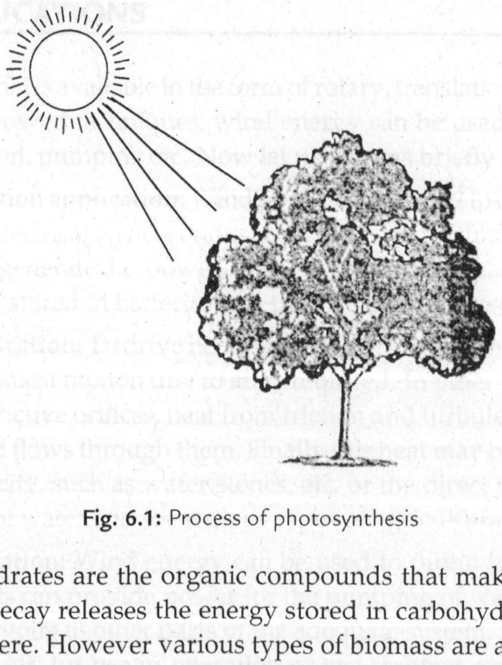

Fig. 6.1: Process of photosynthesis

Thus these carbohydrates are the organic compounds that make up biomass. When these plants die, the process of decay releases the energy stored in carbohydrates and again discharges CO_2 back into the atmosphere. However various types of biomass are as given in Fig. 6.2.

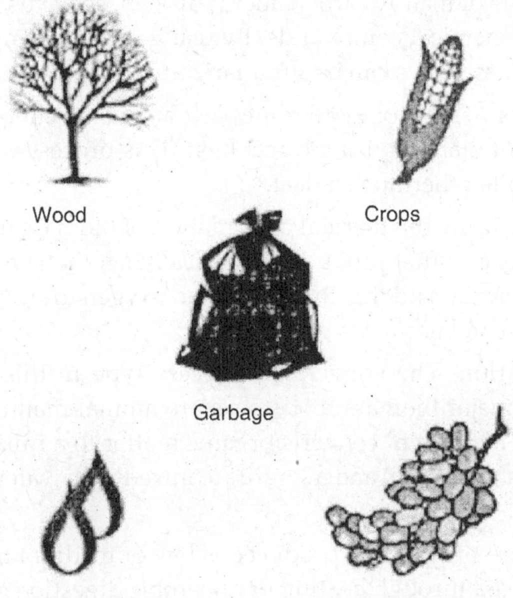

Fig. 6.2: Different types of biomass

Biomass is a renewable energy source, developed or created by carbonaceous waste of various human and natural activities, or in other words it can be said that it is called renewable source, because we are going to carry on making waste products and these waste products are related with biogas, that is environmental friendly source. The biomass energy does not produce CO_2 emissions and avoids the problem of global climate change or in other sense; it is a process to create energy by disposing the waste materials, so that the chances of environmental-risks are avoided.

Biomass can be easily converted to other usable forms of energy like methane gas or transportation fuels like ethanol and biodiesel. Landfill gas (biogas) or methane gas is actually released by rotting garbage, agricultural and human waste. Crops like corn and sugar-cane can be fermented to create the transportation fuel, ethanol. Likewise Biodiesel (other transportation fuel), can be produced from leftover food products like vegetable oils and animal fats.

BIOMASS—THE GROWING ENERGY RESOURCE

We know that biomass is basically a basic term for plants, bacteria, animals etc. and overall earth's biomass represents a very large amount of energy. Energy from biomass is important in such a way that it converts waste things into valuable form. Sun provides the energy, which remains present in biomass. Solar energy in the form of photons is used with air and water so as to develop a number of compounds. These compounds include sugars, starch and cellulose—collectively called carbohydrates. The original energy is now stored in the chemical bonds of these compounds.

This biomass energy source is at its starting position and soon it is going to be tapped. It burns the residual fiber waste from raw sugar processing—called as bagasse so as to produce the steam. Finally this steam is used to drive the alternators, thus producing electricity.

Since biomass consists of mainly carbon and hydrogen, so we use the technology to free the energy, bound up in these chemical compounds. By using some ways, stored chemical energy in biomass can be captured. These ways can be summarized as follows:

(*a*) **Direct combustion.** This is one of the simplest biomass technologies and in general term, it is the burning of material using direct heat. This process is called conversion process and the type is called thermochemical.

(*b*) **Pyrolysis.** It is basically the thermal degradation of biomass using heat in the absence of oxygen. The energy or fuel produced is basically synthetic fuel oil (biocrude) charcoal. In other sense, it can be said that there exists no oxygen to support combustion, therefore creation of fuel oil and gas takes place.

(*c*) **Anaerobic digestion.** The conversion process type in this digestion is biochemical (anaerobic). The major biomass foodstocks are animal manure, agricultural waste and wastewater etc. This system converts organic matter to a mixture of methane. Biomass, such as water waste (sewage) and manure are mixed with water and fed to a tank (digester tank) without air.

(*d*) **Gasification.** The energy or fuel produced is low or medium Btu producer gas. Methane gas can be produced through heating or anaerobic digestion syngas, which is a mixture of CO and H_2, can be easily derived from biomass.

(*e*) **Landfill gas.** It is produced by the decay of buried garbage in landfills. When organic waste decomposes, a gas consisting of nearly 50% methane is generated.

(*f*) **Fermentation of alcohol.** Fuel alcohol is produced by converting starch to sugar, fermenting the sugar to alcohol, then separating the alcohol water mixture by distillation. Likewise foodstock such as wheat, barley, potatoes and waste paper, sawdust and straw containing sugar, starch or cellulose can be converted to alcohol by fermentation with yeast.

(*g*) **Cogeneration.** It is the simultaneous production of more than one form of energy using a single fuel. Since cogeneration produces heat as well as electricity, it can be proved as a source having more potential compared to biomass generation.

COGENERATION AND BIOMASS GASIFICATION PROGRAMMES

(*a*) **Biomass power/Cogeneration programme.** The programme is being implemented for optimum use of country's biomass resources and exploitation of the biomass power generation potential. The main purposes for biomass power/cogeneration programme are as follows:

(*i*) Improvement and modification of a biomass resource atlas for India.

(*ii*) Enlarge the activities through awareness creation, seminars, workshops, business meets etc.

(*iii*) Provide support to research and development activities for various technologies as well as applications research for enhancement of potential in identified areas. This R and D programme aim at upgradation of the existing biomass conversion technologies, improve capacity utilization and develop much more efficient and cost effective new technology routes.

(*b*) **Biomass gasification programme.** It is another important programme. During past years a multi-institutional co-ordinated project on Advanced Biomass gasification (ABG) was successfully completed. Development and testing of HPG (High Pressure Gasifier) coupled with gas turbine engines for generating electricity has been demonstrated with contributions from all the four participating institutions, namely: IIT (Chennai), IICT (Hyderabad), BHEL (Trichi) and IISC (Bangalore).

However the basic arrangement of biomass-gasifier is as given in Fig. 6.3.

Fig. 6.3: Biomass gasifier

In these programmes some important parameters such as gas composition, pressure and temperature were recorded and analyzed. Likewise, control system for governing gas turbine operations was built and tested. Thus it can be said that these types of programmes are continuously running and these will provide a boon for the electricity market.

BIOMASS AND BIOFUELS

Biomass is a raw source for the production of biofuel. This biomass is known as a renewable source due to the rate at which it takes to regenerate. They are considerably less carbon neutral compared to other forms of renewable energy. In another sense, they are excellent alternative to petrol and diesel for transportation (but perhaps not optimal to replace fossil fueled power plants).

Biofuels are basically **first generation fuels,** which are made up from sugar, starch, vegetable oil or animal fats, by using conventional technology. But as the improvement goes on, the next generation fuels come on the top. In relation to this, **second generation fuels** are made from lignocellulosic biomass feedstock using advanced technical processes: Biohydrogen, Biomethanol, Fischer-Tropsch diesel and mixed Alcohols. Actually these biofuels contain no petroleum, but to produce a biofuel blend, it can be easily blended at any level with petroleum fuel. It can be used in

Thus overall, government has much more planning for electricity generation using biomass. With the assistance of private sectors, it can be hoped that in near future, new units with above technology will be established and this source will be proved as a better source for generation in our India.

BIOMASS AND THE ENVIRONMENT

Biomass is a source, directly concerned with our environment. This source absorbs CO_2 during growth and emits it during combustion, so it recycles atmospheric carbon and does not produce greenhouse effect. Burning biomass fuels doesn't produce pollutants like sulphur that can cause acid rain. One important fact regarding biomass is that, when biomass crops are grown, a nearly equivalent amount of CO_2 is captured through photosynthesis process. Thus overall it can be said that each of the forms and uses of biomass impacts the environment in a different way. For the better understanding, the carbon cycle may be considered as given in Fig. 6.4.

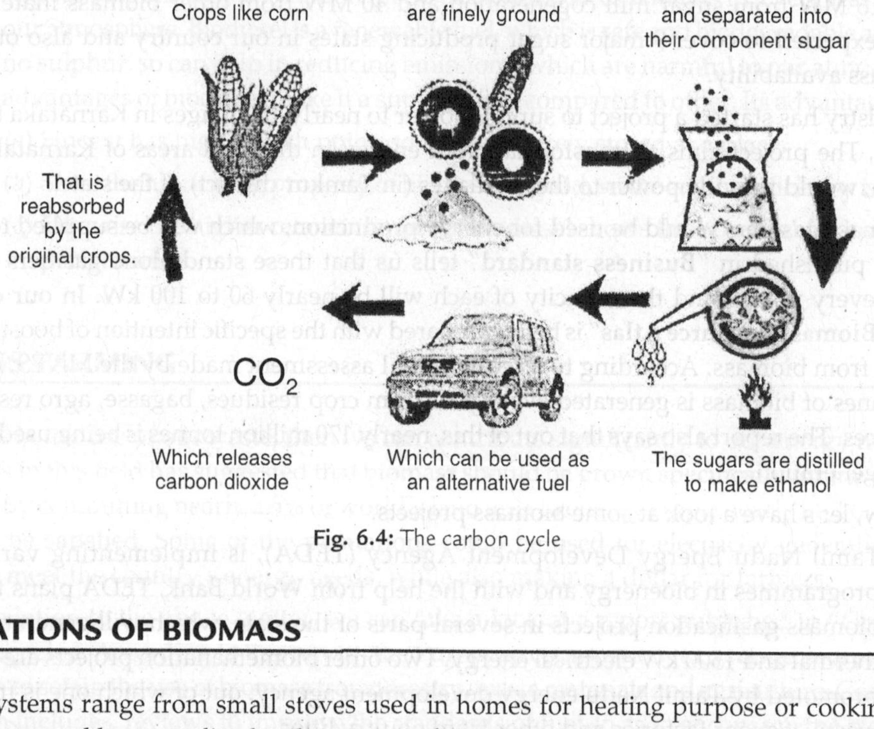

Fig. 6.4: The carbon cycle

APPLICATIONS OF BIOMASS

Biomass systems range from small stoves used in homes for heating purpose or cooking to large power plants used by centralized utilities to produce electrical power.

(*i*) **Residential application.** It can be used for heating purpose mainly in homes and for this, one of the most important source for fuel is wood. However when new designs are used, then the efficiency of these systems can be increased by reducing the quantity of fuel needed.

(*ii*) **Power generation application.** In this application, the mechanical energy produced will be utilized to drive alternators coupled to it and thus generation of electrical power takes place. The power may be used for either local consumption or for grid synchronization.

The power plant can generate power at around 2.5 rupees per unit (kWHr) excluding investment interest.

(*iii*) **Thermal application.** Like any other gaseous fuel, producer gas has finer power control capability. The flame temperatures as high as 1200°C can be gained by optimal pre-mixing of air with producer gas. The gasifiers may be a good option as a gas generator, which needs thermal energy.

(*iv*) **Industrial application.** Industries use biomass for various purposes. It may be used for space heating, hot water heating and for electricity generation (as discussed above). It is hoped that by the year 2014, the applications are expected to be triple as biomass technologies are continuously rising one.

ADVANTAGES AND DISADVANTAGES

(*a*) **Advantages:**

(*i*) It provides a sustainable, affordable and eco-friendly alternative to fossil fuel.

(*ii*) Waste materials can be utilized, so in this sense it is of great importance.

(*iii*) Encourage energy plantation, thus resulting in a green environment.

(*iv*) Improvement of nation's energy security using locally available fuels.

(*v*) Can be combined with production of other useful products, making it an attractive fuel.

(*vi*) Biomass-based generation systems make a link to plantations on wasteland and simultaneously address the vital issues of wasteland development, environmental restoration rural employment generation and generation of power with no distribution losses.

(*vii*) Lower cost of production and simplicity in operation.

(*b*) **Disadvantages:**

(*i*) Producing the waste in sufficient quantities, may be difficult.

(*ii*) During indoor cooking and in the absence of sufficient ventilation fuels, such as dung cause air pollution, which becomes a serious problem.

Chapter 7

FUEL CELLS

INTRODUCTION

A fuel cell is a device in which the chemical energy is converted directly into electrical energy. One of the important aspects of the fuel is that the fuel and its oxidants are mixed in the form of ions rather than neutral molecules. First practical fuel cell was designed and developed by scientists, namely Francis T. Bacin and J.C. Frost of Cambridge University. These fuel cells are being proved as an important source of power generation in various countries including U.S.A. These cells will be beneficial for supply of electrical energy in rural villages and remote areas. The important benefit from this cell is the saving of money, because this source will not require transmission lines.

Amongst the fuels used in fuel cells, hydrogen is very important. This is due to the fact that hydrogen and oxygen are capable of releasing more energy per unit weight than most other oxidizer combinations. The other fuels are hydrocarbons, Ammonia etc. Hydrocarbons are basically methane, ethane etc. These are less reactive than hydrogen and it becomes very tedious task to oxidize them. Similarly Ammonia is not used generally due to its low reactivity and finally the choice of fuel, for a Fuel cell is governed by availability, transportability and volume etc. These fuel cells are free from air pollution, lighter in weight, safer and have larger raw material availability. However the basic arrangement of fuel-cell is as given in Fig. 7.1.

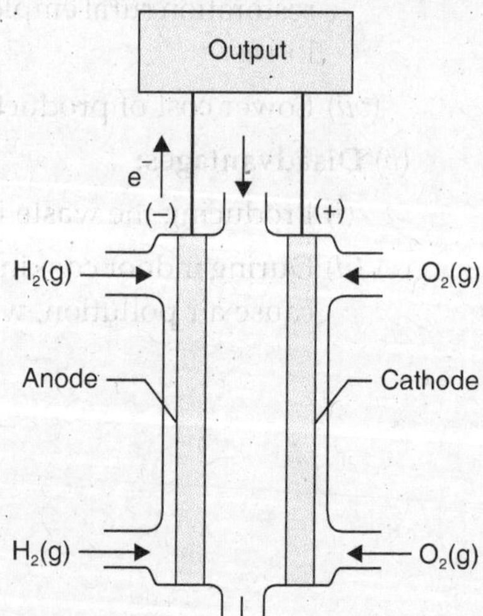

Fig. 7.1: Arrangement of fuel-cell

As these fuel cells produce power without combustion, they are more efficient. The feature, which makes this cell an attractive source, is very high energy density of liquid fuels (such as gasoline, methanol etc.) and compressed hydrogen.

WORKING OF FUEL CELL

A fuel cell consists of an anode (positive end of circuit) and a cathode (negative end of circuit) separated by an electrode. Fuel flows to the anode and an oxidant flows to the cathode. Thus finally electricity is produced due to the chemical reaction between, the fuel and the oxidant. The basic block diagram is as shown in Fig. 7.2.

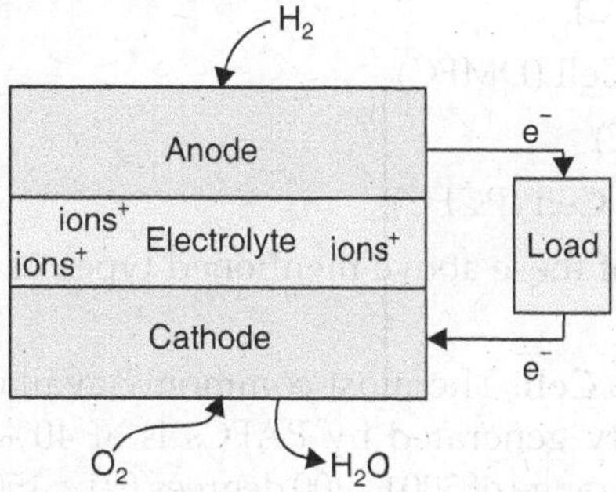

Fig. 7.2: Basic block diagram of fuel-cell

A number of fuel cells are used, but one of the important cell is hydrogen fuel cell. In the hydrogen fuel cell, hydrogen is supplied to the fuel cell's anode and an oxidant, commonly the oxygen is supplied to the cathode. The fuel cell strips electrons from the hydrogen atoms and finally electrical current is developed due to the movement of electrons from anode to cathode through the complete circuitry. Meanwhile the electron deficient hydrogen atoms pass through the electrolyte to the cathode. There electrons (that passed through the circuit) recombine with electron deficient hydrogen atoms. Oxygen from the air reacts with this reformed hydrogen, producing water. Here an important precaution must be taken that water (which is produced at the cathode) has to be removed continuously to avoid the flooding of the cell. However Fig. 7.3 clearly shows the working of fuel-cell.

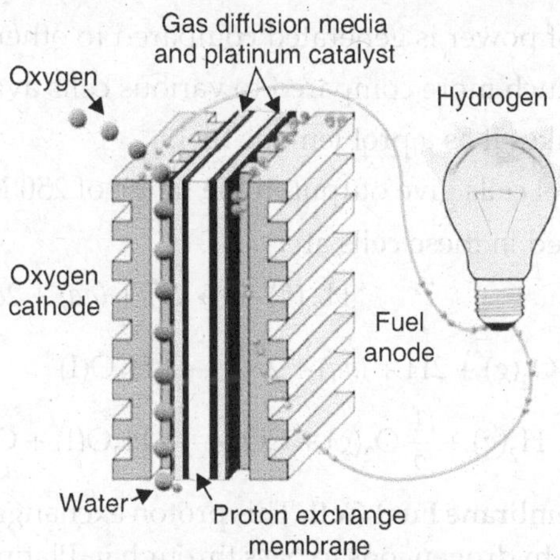

Fig. 7.3: Working of fuel-cell

TYPES OF FUEL CELL

Fuel cells are generally categorized by their electrolyte and the material sand witched between the two electrodes. Material's characteristics determine the optimal operating temperature and the fuel used, to generate electricity. The various types of fuel cells, which are of great importance are as follows:

(*i*) Phosphoric Acid Fuel Cell (PAFC)
(*ii*) Proton Exchange Membrane (PEM) or Solid Polymer Fuel Cell
(*iii*) Molten Carbonate Fuel Cell (MCFC)
(*iv*) Solid oxide Fuel Cell (SOFC)
(*v*) Alkaline Fuel Cell (AFC)
(*vi*) Direct methanol Fuel Cell (DMFC)
(*vii*) Zinc air Fuel Cell (ZFC)
(*viii*) Protonic Ceramic Fuel Cell (PCFC).

Now let us discuss each of these above mentioned types of fuel cells. The brief description is as follows:

(*a*) **Phosphoric Acid Fuel Cell.** The most commonly available fuel cell is phosphoric acid fuel cell. The electricity generated by PAFCs is at 40% efficiency rate. The operating temperatures are in the range of 300 to 400 degrees F (*i.e.* 150–200 degrees C). One important thing regarding this cell is that at lower temperatures, phosphoric acid becomes a poor ionic conductor and poisoning of the platinum (Pt) electro catalyst in the anode becomes severe.

(*i*) **Advantages:**

- Nearly 85% cogeneration efficiency is obtained.
- It can use impure hydrogen as fuel.
- Carbon monoxide concentration of nearly 1.5% can be tolerated by this cell, which actually broadens the choice of fuels, which they can use.

(*ii*) **Disadvantages:**

- Platinum is used as a catalyst, which is really very expensive.
- Low amount of power is generated compared to other cells.
- Its weight is much more compared to various cells available.
- Larger size makes it as a problematic fuel.

(*iii*) **Output:** These fuel cells give output in the range of 250 MW maximum.

Equations involved in these cells are:

At anode: $$H_2\,(g) \longrightarrow 2H + (aq) + 2e^-$$

At cathode: $$\frac{1}{2}O_2(g) + 2H + (aq) + 2e^- \longrightarrow H_2O(I)$$

Cell: $$H_2(g) + \frac{1}{2}O_2(g) + CO_2 \longrightarrow H_2O(I) + CO_2$$

(*b*) **Proton Exchange Membrane Fuel Cell.** This proton exchange membrane is a thin sheet of plastic, which allows hydrogen ions to pass through it. Platinum (highly dispersed metal alloy particles) is used for coating purpose on both the sides of membrane. Solid organic polymer, poly-perflourosulfonic acid is used as an electrolyte. Hydrogen is fed to the anode side of the fuel cell where the catalyst encourages the hydrogen atoms to release electrons and finally they become hydrogen ions (protons). The electrons move in the form of current then can be used before it returns to the cathode side of fuel cell, where oxygen has been fed. At the same moment, the protons diffuse through the membrane (electrolyte) to the cathode, where hydrogen atom is recombined and reacted with O_2 so as to produce water thus completing the total process. These cells can be operated at about 175 degrees F or 80 degrees centigrade.

(*i*) **Advantages:**

- These possesses high value of power density and the output can be varied in a quick manner, as according to demand.

(*ii*) **Disadvantages:**

- It is sensitive to fuel impurities, so makes an undesirable factor.

(*iii*) **Output:** The fuel cell output is in the range of 50 to 250 kW.

Equations involved in these cells are:

At anode: $H_2(g) \longrightarrow 2H + (aq) + 2e^-$

At cathode: $\frac{1}{2} O_2(g) + 2H + (aq) + 2e^- \longrightarrow H_2O(l)$

Cell: $H_2(g) + \frac{1}{2} O_2(g) \longrightarrow H_2O(l)$

(*c*) **Molten Carbonate Fuel Cell (MCFC).** A liquid solution of lithium, sodium and/or potassium carbonates, soaked in a matrix for an electrolyte is used in these cells. They operate nearly at 1200 degrees F or 650 degrees centigrade. So as to get sufficient conductivity of electrolyte, high value of operating temperature is required and due to this high temperature, nobal metal catalysts are not needed for the electrochemical oxidation and reduction processes of the cell. Hydrogen, natural gas, marine diesel, carbon monoxide, propane and landfill gases have been used for molten carbonate fuel cell.

(*i*) **Advantages:**

- High value of fuel to electricity efficiency, nearly 85%, with cogeneration is obtained.
- Higher operating temperature provides flexibility to use more types of fuels.
- High value of efficiency.

(*ii*) **Disadvantages:**

- Due to the high value of temperature, corrosion may take place. It may also lead to breakdown of cell components.

At anode: $H_2(g) + CO_{32} \longrightarrow H_2O(g) + CO_2(g) + 2e^-$

At cathode: $\frac{1}{2} O_2(g) + CO_2(g) + 2e^- \longrightarrow CO_{32}^-$

Cell: $H_2(g) + \frac{1}{2} O_2(g) + CO_2(g) \longrightarrow H_2O(g) + CO_2(g)$

(*d*) **Solid Oxide Fuel Cell (SOFC).** This fuel cell consists of a solid oxide system with hard ceramic material of solid zirconium oxide and a very small quantity of ytrria (instead of a liquid electrolyte). The operating temperature for this cell is 1800 degrees F and 1000 degrees centigrade approximately. The efficiencies are nearly 60% and 85% with cogeneration.

(*i*) **Advantages:**

- Can be easily used for industrial and large scale central electricity generation systems.

(*ii*) **Disadvantages:**

- Overall cost of the system may be quite high for large systems.

(*iii*) **Output.** The output may vary in the range of 25kW–220 kW.

The equations involved in the whole process is as follows:

At anode: $H_2(g) + O_2 \longrightarrow H_2O(g) + 2e^-$

At cathode: $\frac{1}{2} O_2(g) + e^- \longrightarrow O_2^-$

Cell: $H_2(g) + \frac{1}{2} O_2(g) \longrightarrow H_2O\ (g)$

(*e*) **Alkaline Fuel Cell (AFC).** Sometimes alkaline cells are used, for some specific purposes. In this cell an aqueous solution of alkaline potassium hydroxide soaked in a matrix as the electrolyte, is used. Some companies are using such kind of cells nowadays. The operating temperature is in the range of 150–200 degrees C (or nearly 300 to 400°F).

(*i*) **Advantages:**

- Power generating efficiency up to nearly 70% can be achieved.
- Higher performance is achieved by these cells, because cathode reaction is faster in the alkaline electrolyte.

(*ii*) **Disadvantages:**

- Main hurdle in the path of AFC is its cost. Actually it is too costly for commercial applications.

(*iii*) **Output.** The output is in the range of 300 W to 5 kW.

The equations involved in the process are:

At anode: $H_2(g) + 2(OH)^-\ (aq) \longrightarrow 2H_2O(l) + 2e^-$

At cathode: $\frac{1}{2} O_2(g) + H_2O(l) + 2e^- \longrightarrow 2(OH)^-\ (aq)$

Cell: $H_2(g) + \frac{1}{2} O_2(g) \longrightarrow H_2O(l)$

(*f*) **Direct Methanol Fuel Cell (DMFC).** These cells are same as PEM cells, because both of them use a polymer membrane as the electrolyte. In these cells, hydrogen from the liquid methanol is drawn by anode catalyst itself. The efficiency obtained is nearly 40%. The temperature remains in the range of 120–190 degrees F or 50–100 degree C.

(*i*) **Advantages:**

- Since operating temperature range is low, so this cell can be used for small and medium size applications (for example in laptops and cellular phones).
- Higher values of efficiencies are obtained at higher temperature.

(*ii*) **Disadvantages:**

- The main problem is the fuel crossing over from the anode to the cathode without producing electricity.

The equations involved are as given below.

At anode: $CH_3OH\,(aq) + H_2O\,(l) \longrightarrow CO_2(g) + 6H + (aq) + 6e^-$

At cathode: $6H + (aq) + 6e^- + \frac{3}{2}O_2(g) \longrightarrow 3H_2O(l)$

Cell: $CH_3OH\,(aq) + \frac{3}{2}O_2(g) \longrightarrow CO_2(g) + 2H_2O(l)$

(*g*) **Zinc Air Fuel Cell (ZAFC).** This type of cell consists of a zinc anode separated by electrolyte. Some form of mechanical separators and a gas diffusion electrode. The oxygen in atmosphere is allowed to pass through gas diffusion electrode. After the conversion of oxygen into hydroxyl ions and water, these ions move through electrolyte and reaches the zinc anode. At this situation, it reacts with the zinc and forms zinc oxide. Finally the electrical current develops causing an electric potential.

(*i*) **Advantages.** It has high specific energy, which is an important factor in determining the running duration of a battery relative to its weight. Since zinc is available easily on earth, so the overall cost for ZAFCs and zinc-air batteries are low.

(*ii*) **Disadvantages.** In these cells, refueling is very different and difficult and shares characteristics with batteries.

(*h*) **Protonic Ceramic Fuel Cell (PCFC).** It is based on a ceramic electrolyte material, which shows high protonic conductivity at elevated temperatures. For these cells, high operating temperature is required, because with hydrocarbon fuels, very high electrical fuel efficiency can be obtained. On the surface of anode, gaseous molecules of hydrocarbon fuel are absorbed in the water vapour presence and hydrogen atoms are stripped off to be absorbed into the electrolyte. In the complete process, CO_2 becomes primary reaction product.

(*i*) **Advantages:**

- The process reduces the chances of intermediate step of producing hydrogen.
- These cells consist a solid electrolyte so the membranes will not dry out and the liquid will not leak out.

(*ii*) **Disadvantages.**

- The total cost is high compared to some other types of cells.

IMPROVEMENT IN THE FUEL CELL USING NANOTECHNOLOGY

Nanotechnology is basically about making extraordinary devices with normal matter. In other words, it can be said that it is a very vast field, a super set of physics, biology, engineering and lot of other fields of science. This nanotechnology directly impacts the fuel cell. In these cells, a number of catalysts are used such as hydrogen, methanol, platinum etc. Out of these catalysts, one of the important catalysts is platinum. Nowadays, nanoparticles of platinum are being preferred because it reduces the amount of Pt needed. Some improved techniques in nanotechnology completely replace the platinum, thereby reducing the overall cost of the system.

These fuel cells have membranes that allow hydrogen ions but do not allow other atoms or ions, such as oxygen to pass through them. By using nanotechnology, more efficient membranes may be designed, thus giving the lighter weight and longer lasting fuel cells. The fuel cells, which can be used to replace batteries in electric cars, are also under development processes using nanotechnology. Since hydrogen is the most commonly used fuel, so it becomes necessary to develop a light weight and safe hydrogen fuel tank so as to hold the fuel. Researchers are developing light weight nanomaterials that will absorb the hydrogen and only release it, when needed.

The electrolyte of PEM fuel cell can be improved by nanoparticles. For SOFC (solid oxide fuel cells), ceramic powders such as yttrium, stabilized zirconium are used for producing solid electrolyte membranes, with improved ionic conductivity and good thermal stability. Likewise nanomaterials have the potential to be a light weight, high efficient storage media for hydrogen due to their increased active surface area. Thus in brief, it can be said that by using nanotechnology improved devices with high value of efficiency can be obtained.

IMPROVED FUEL CELL CATALYST TECHNOLOGY

Nowadays, catalyst plays an important role in the development of fuel cells. If however cheaper catalyst can be produced, then it will be beneficial for the alternative energy technology. Similarly, if we take the case of platinum, then we find that it is very much expensive. Recently a corporation at California, known as "**Quantum Sphere Inc**", has designed a nickel-cobalt nanomaterial that could replace platinum catalyst in a variety of batteries and fuel cell applications. In brief, it can be said that the net cost of the system would be reduced by nearly 50%.

However if nickel-cobalt nanomaterial is used, then some sacrifice in terms of performance must be done. In fact, if all platinum on the cathode is replaced by nanomaterial made up of nickel-cobalt, then the total cost will reduce by 90% and performance would decline 27% compared with pure platinum. The above mentioned company *i.e.* Quantum Sphere is following up the preliminary results with a series of optimization experiments. At present the alloy consists of nearly 80% nickel and 20% cobalt, but these compositions may be varied as according to the requirement of system. The scientists are continuously performing experiments related to these matters. They grow a uniform distribution of required particle size. Inside a vacuum chamber, gas phase condensation takes place, where metal wire is melted at high temperatures. The metal vapour is collected and inert gas is used for cooling purpose. Now due to condensation, it is converted into droplets of liquid metal. By varying the temperature and pressure of the chamber, the particle's characteristic can be changed.

In the continuation of various successful experiments, "**Hitachi Maxwell Ltd.**" has developed a technology, which creates very fine catalysts that could reduce the net fuel cell cost. Actually phosphorous is mixed with platinum ruthenium, which is mostly used as a catalyst in fuel cells. Due to the production of finer catalyst, the efficiency and the reactive surface area are increased. By tests, it was also reported that these new catalysts obtained a maximum output density of nearly 1.7 times the conventional platinum ruthenium catalysts.

So finally, the result is that the use of platinum can be cutdown and new technology may be used, which reduces the manufacturing cost of the system.

APPLICATIONS OF FUEL CELL

The fuel cell is becoming popular as an energy source. It can have application in number of areas, such as electrical power production, residential, stationary and transportation applications, portable/ backup power applications, military applications etc. Now let us have a brief introduction of all these applications.

(*a*) **Electrical power production.** Fuel cells are most ideal for electrical power production. Although in today's world, batteries are being used for a no. of devices, but they do not have long life. Due to this fact, attraction goes towards fuel cell, because fuel cells could provide continuous power. In future, it will be a great technology for small electric power plants. They possesses good efficiencies and run quietly without any disturbance. They emit very little pollution and requires less maintenance. Thus all these factors, make it a useful device for electricity production.

(*b*) **Residential application.** These cells find applications for power generation either connected to the electric grid to provide supplemental power and backup assurance for critical areas, or installed as a grid-independent generator for onsite service in areas that are inaccessible by power lines. These cells produce heat, which can be used to provide hot water or space heating purpose at home.

(*c*) **Stationary application.** It can also be used for stationary applications. They can be used in hotels, office buildings, schools, utility power plants, nursing homes and in hospitals also.

(*d*) **Transportation application.** These cells can be used to replace Otto or diesel engines. Nowadays fuel cell powered cars are being preferred in some cities like Japan and California. This fuel car will be like an electric car but with a fuel cell and reformer instead of batteries. Automaker companies hopes that these vehicles will be commercialized by 2014. In brief, we can say that these cells are being incorporated into buses, locomotives, airplanes and scooters.

(*e*) **Portable/buckup power application.** Miniature fuel cells will bring revolution in the telecommunication world. By using these cells, consumers will talk for up to a month on a cellular phone without recharging. These small (micro) fuel cells will be used in video recorders, portable power tools, hearing aids, smoke detectors, meter readers, hotel locks etc.

(*f*) **Military application.** These cells are also beneficial for military applications. These energy sources save money and reduce the pollution at military installations. Mainly they save lives and material by reducing heat and noise. It also has been found that significant war-fightening, logistics and cost, benefits occur, when weapon-systems are made more fuel-efficient.

(*g*) **Wastewater treatment.** These fuel cells operate at landfills and wastewater treatment plants all over the country. Thus it reduces the emission and the power is generated from the methane gas, they produce.

ADVANTAGES OF FUEL CELL

A number of benefits are provided by the fuel cells. All these advantages are as follows:

(*a*) **High reliability.** It is one of the best benefits of fuel cells. They are providing completely grid independent power and sometimes grid can be used as a backup system. By installing a number of units, greater amount of electricity can be obtained, giving high reliability nearly 99.98%.

(*b*) **High efficiency.** These fuel cells do not burn the fuel, so they possesses somewhat greater efficiencies compared to combustion systems.

(*c*) **Security of supply.** These cells can enable a transition to a secure renewable energy future, based on the use of hydrogen. These fuel cells can be combined with various renewable technologies, such as solar, wind etc., to have an energy system with zero emission that needs no fossil fuel. In other sense, they may be used as a continuous and reliable supply source.

(*d*) **Reduction in greenhouse gases.** Due to the use of fossil fuels, greenhouse gases are created. These gases trap the heat and thus warm the earth, because they prevent a significant portion of infrared radiation from escaping into space. But by using these fuel cells, CO_2 emissions can be reduced up to a great level.

(*e*) **Environmental safety.** Hydrogen gas is basically nontoxic and safe to breathe. Air pollution is an important factor regarding environment but the use of fuel cells can reduce pollution and can offer the promise of eliminating pollution.

(*f*) **Smooth operational feature.** It has been proved as a more pleasant operation for transit riders. Its quieter operation makes it suitable for various applications.

(*g*) **International benefits.** In today's world, countries are facing problems of power shortage and thus the challenge for fuel cell industry is to ensure that it is always ready with competitive market. It will be soon proved as an energy technology that will provide greater sustainability in resource consumption.

Other benefits include broad versatility cogeneration, modularity, distributed generation, high quality power, fast response to load variations and engineering simplicity.

LIMITATIONS OF FUEL CELL

Although fuel cell is having a number of advantages, but it has certain limitations also. It can be divided into two main parts namely **internal** and **external** limitations.

(*a*) **Internal limitation.** In fuel cells, two internal factors such as voltage drop and internal resistance create problems. In these cells, some of the energy is lost due to the over potential of anodic and cathodic reactions and finally dramatic voltage drop takes place. So as to reduce this drop in voltage, more efficient materials are required to be used.

Similarly internal resistance affects the fuel cell performance, because it reduces the efficiency of the cell due to the internal voltage drop.

(*b*) **External limitation.** It includes hydrogen availability, pollution factor, leakage of hydrogen, storage of hydrogen and water management.

(*i*) ***Hydrogen availability.*** In fuel cells, two main components are oxygen and hydrogen. Air can be easily obtained from the hydrogen, but it is difficult to have sufficient amount of hydrogen. Hydrogen may be generated by using a reformer, but the development of reformer is not completed yet and also it can not be installed practically. In a broader sense, we can say that the availability of pure hydrogen is difficult.

(*ii*) ***Pollution factor.*** Hydrogen gas is required to be purified and during the purifying process, carbon dioxide's production takes place, so it will create little amount of harmful substances.

(*iii*) ***Leakage of hydrogen.*** It means disturbance. Leakage of hydrogen produces a disturbance in the ozone chemistry. The actual problem is that as the hydrogen goes up into the stratosphere, it would condensate and oxidize, so it would create problems.

(*iv*) ***Storage of hydrogen.*** The net cost of storing hydrogen is high, which makes it a limitation of fuel cell. In addition to this, with the compressed hydrogen fuel, the distance that a vehicle can travel, is still not desirable.

(*v*) ***Water management.*** In fuel cells, water inside the cell boils at 100°C. If the atmospheric temperature automatically increases, then water will evaporate and result in a dry membrane, which will eventually loose its conductivity. This in this sense, this problem becomes an important issue.

Chapter 8 MISCELLANEOUS

1. GEOTHERMAL ENERGY

- **Introduction**

Geothermal energy is one of the renewable energy sources, which is based on the natural energy flows of the earth. The word geothermal is the combination of Greek word **geo (earth)** and **therme (heat)**, hence the energy is called geothermal energy *i.e.* heat from within the earth. This source is called as a renewable energy source, because the water is replenished by rainfall and heat is continuously produced inside the earth's surface or in other words, we can say that this energy will continue in the process of cooling for the indefinite future, as it is a renewable energy source.

This energy is generated in the core of the earth, which is nearly 4000 miles below the surface of earth. Due to the decay of radioactive particles, the temperature inside the surface becomes very high and hence becomes a source of power generation. The layers of earth are called core, mantle and crust. The **core** has basically two layers namely solid iron core and magma (outer core made of very hot melted rock). **Mantle** is about 1800 miles in thickness and made up of rock and magma. Likewise outermost layer of earth is called **crust**. The thickness is nearly 15 to 35 miles on the continents.

This energy source is used for heat and power generation purpose and due to this fact, a significant portion of demand is being covered by this source in some developing nations.

These plants may be in the range of 100 kW to 100 MW depending on the demand of power. In those nations, where there are no fossil fuel resources, (Such as coal, oil, natural gas), this energy source may work wonders. However it is being expected that India's first geothermal power plant with an initial capacity of 25 megawatts will be coming up in Andhra-Pradesh's Khamman district by 2014.

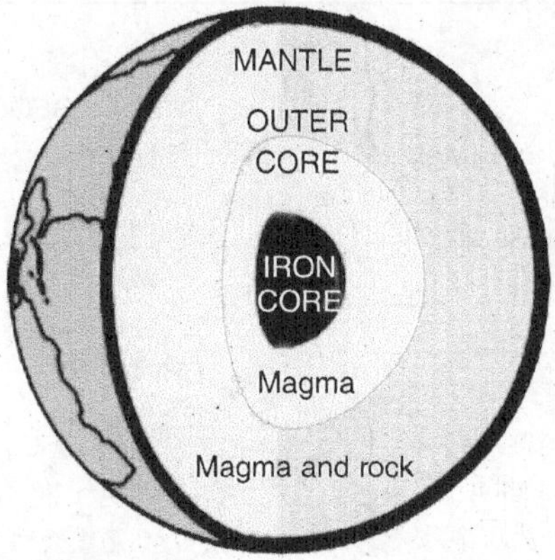

Fig. 8.1: Earth's interior portion

- **Suitable Places for Geothermal Energy**

Geothermal power is generated all around the world in more than 15 countries, like Iceland, Africa, New Zealand, Costa Rica, Philippines, Indonesia, China, Japan, Saint kitts and Newis, Mexico, Russia, France, Italy, United states, Australia etc. All these above mentioned nations are largest producers of this energy in the world. In this context, geothermal power use by some countries (Five top most countries) is as given in Table 8.1.

Table 8.1: Geothermal Power use by Countries

S. No.	*Country*	*% Use*
1	China	19.5%
2	United states	12.6%
3	Russia	3.8%
4	Japan	3.5%
5	France	3%

In fact a number of geothermal reservoirs are deep underground with no visible clues. This energy remains in the form of volcanoes and fumaroles (holes, where volcanic gases are released). Maximum geothermal activity exists in an area, known as **"Ring of fire."** Some hydrothermal resources consist of two basic components called as water (hydro) and heat (thermal). These large areas are called as geothermal reservoirs. Geologists drill a well and then test the temperature deep underground. If it is up to satisfactory limit, then it is declared as a geothermal reservoir. Most of the geothermal reservoirs in US are situated in the western states. However various provinces for geothermal energy in India are as given in Fig. 8.2.

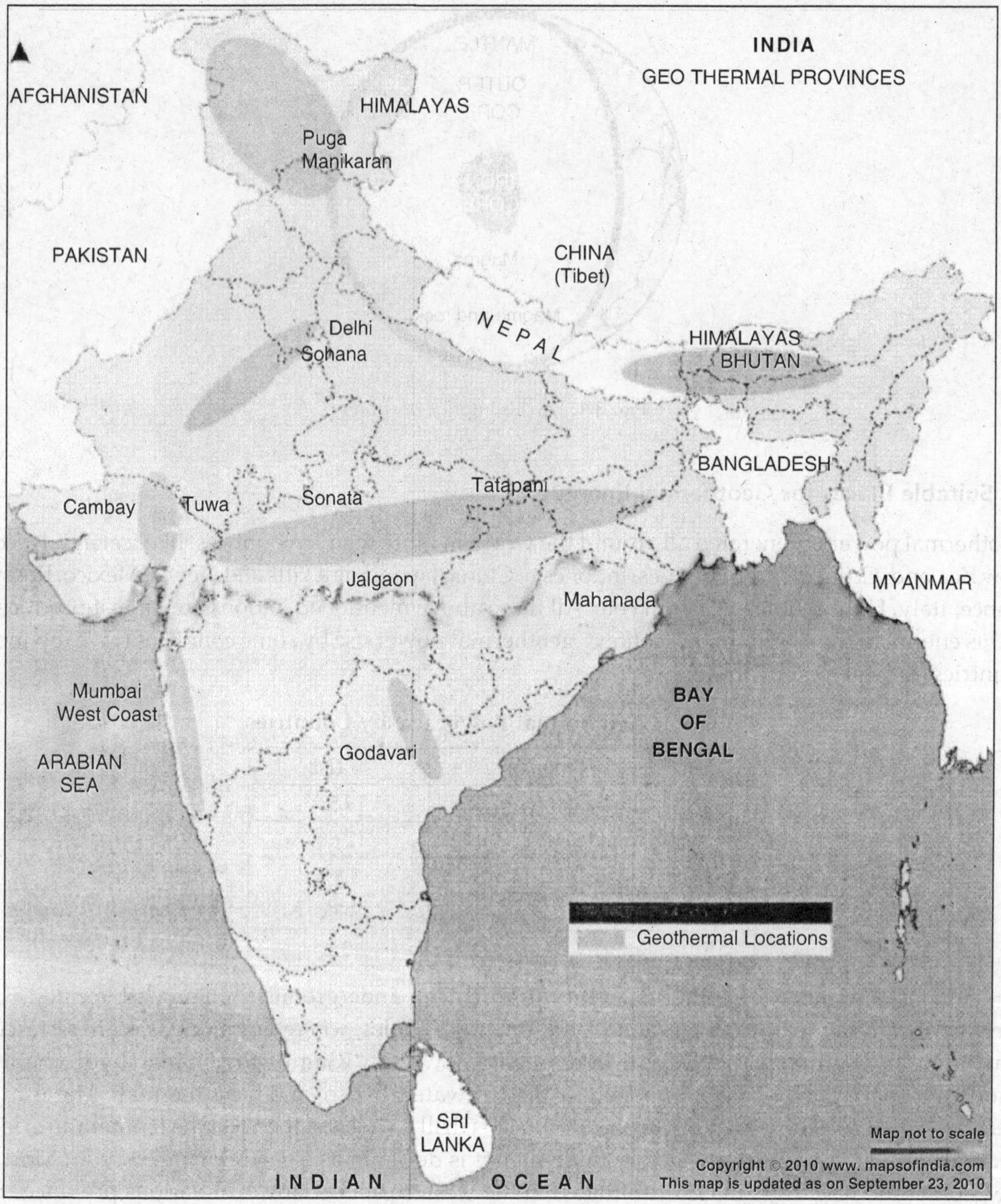

Fig. 8.2: Geothermal provinces in India

- **Geothermal Power Plants**

These geothermal plants are very high temperature (in the range of 300 to 700 degrees Fahrenheit) hydrothermal resources, that may be obtained from either dry steam wells or hot water wells , which are generally 1 to 2 miles deep. United States is a country that generates a large amount of electricity from geothermal resources compared to any other country and consists of only four states having geothermal plants, namely California, Nevada, Hawaii and Utab. These states have 33, 15, 1 and 1 geothermal power plants respectively. The breakdown of geothermal electricity production in various countries may conveniently be represented by pie-chart as given in Fig. 8.3.

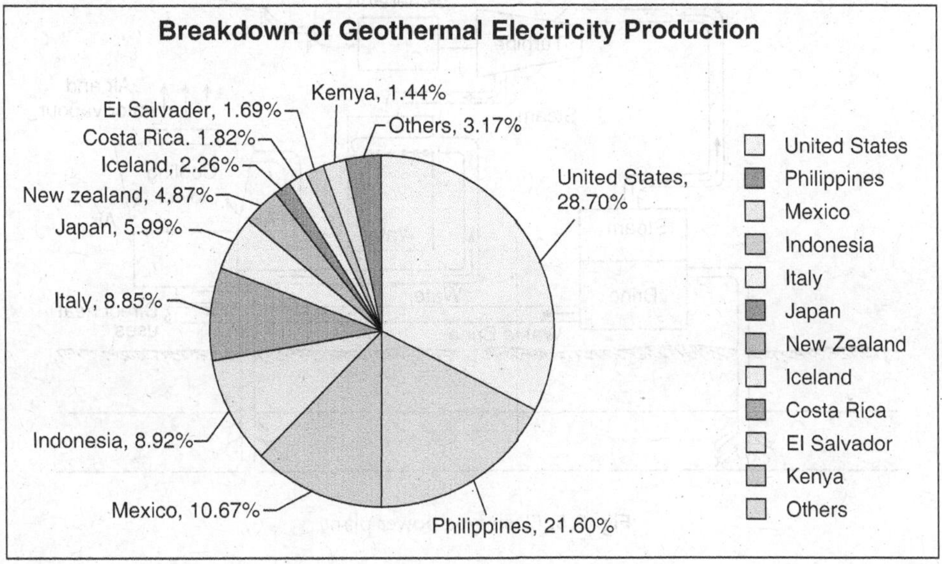

Fig. 8.3: Geothermal energy around the world

There are basically four types of geothermal energy systems, namely:-

(*i*) Dry steam systems/plants

(*ii*) Wet steam systems/Flash steam plants

(*iii*) Hot dry rock systems and

(*iv*) Magmatic (molten rock) chamber systems

Now let us discuss about all these systems briefly.

(*a*) **Dry steam systems.** In these systems, steam from the geothermal field is fed directly to the turbine. By using centrifugal separators, liquid and suspended solid particles are removed. Now a pump is used to feed back the condensate to the hot field.

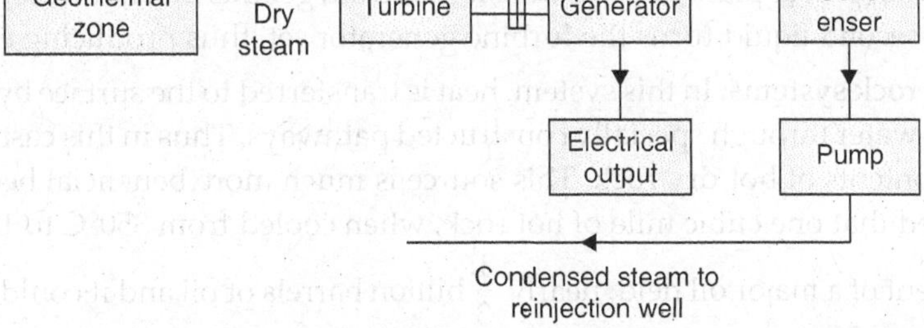

Fig. 8.4: Dry steam systems

(*b*) **Flash steam plants.** In these systems, water coming out from hot well is flashed in flash chamber to produce the steam and finally steam drives the turbine. The cooled steam condensates to water and injected back into the ground to be used over and over again.

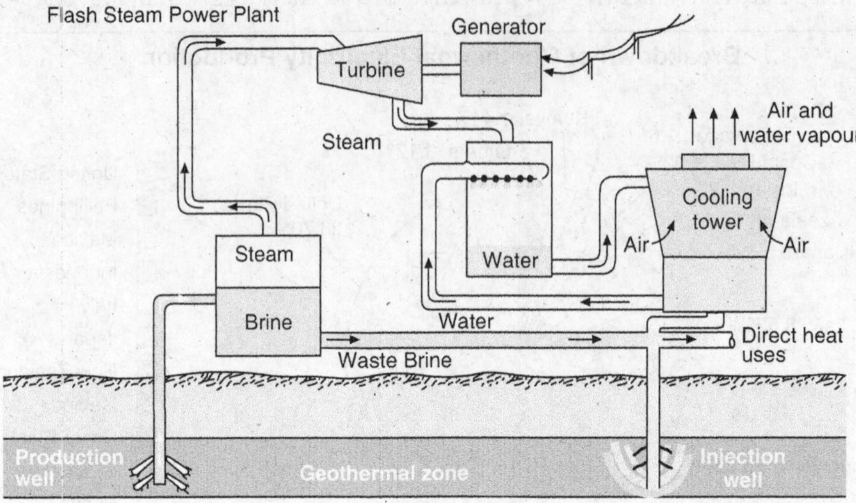

Fig. 8.5: Flash type power plant

Many times the pressure and temperature of water are not sufficient to create the flash steam and in this condition, the heat in the water is used in a closed system.

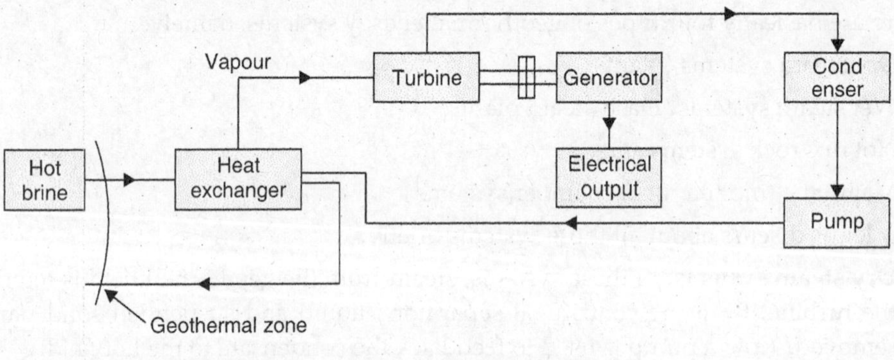

Fig. 8.6: Hot water closed (Binary) system

In this case (power plant), heat is transferred from geothermal hot water to another liquid and the second liquid turns the turbine generator set, thus producing electricity.

(*c*) **Hot dry rock systems.** In this system, heat is transferred to the surface by using artificially injected water through specially constructed pathways. Thus in this case, energy is found in the contents of hot dry rock. This source is much more beneficial because it has been estimated that one cubic mile of hot rock, when cooled from 350°C to 180°C, has energy equivalent of a major oil field, nearly $\frac{1}{3}$ billion barrels of oil and it could yield about 8000 MW-Year of energy, if it could be efficiently recovered.

(*d*) **Magmatic (molten rock)-chamber systems.** At present, the case of harnessing heat energy by hot magma at a temperature of 1450 degree Celsius is under study process. Based on the study results, it is hoped that, an establishment of 100 megawatt power plant would need nearly 400 m^2 of heat exchanger surface area, using an average value of 250 kW/m^2 heat extraction rate. Although at present, there are many problems regarding harnessing of geothermal energy from this system, but in future it is hoped that this system will provide a significant portion of power for various countries.

- **Some Geothermal Issues**

(*a*) **Meeting the demand.** Geothermal plants are suitable for those systems, which would have a great dependency on fossil fuels. These plants are established, so as to meet the growing base load demand and the backup power can be supplied by thermal plants.

(*b*) **Environmental factor.** Environment is directly linked with geothermal plants. These plants do not burn the fuel so as to generate electrical power, that's why the lower CO_2 emissions takes place compared to fossil fuels. An estimation shows that nearly 2–5 % of emissions are produced, which does not affect the atmospheric conditions so badly. These plants use scrubber systems to clean the air of hydrogen sulfide that is naturally found in the steam and hot water. In other words, we can say that energy harnessed is clean and safe for the surrounding environment.

(*c*) **Financial and economic feasibility.** These geothermal stations involve high capital investment. This high investment is needed for drilling the wells, plant installations and exploration purposes etc. Although we cannot get return on our investment in early years but after a long time period, economic benefits may occur.

(*d*) **Sociological issue.** Local and regional communities can be benefitted by this energy source, with minimum impact on existing land uses. Greater benefit can be returned locally in those countries where landowners have some control over access to the geothermal resource.

- **Use of Geothermal Energy**

Since ancient times, hot water is being used as an energy source. It may be used for heating, cooking and bathing purposes. One of the most common uses is for heating the buildings by using district heating systems. The direct use of this energy source is being done in some countries like Romania, China, Algeria etc. In Romania, the direct use of this energy includes sanitary hot water for dwellings, preparation of industrial hot water for various factories, etc. Similarly in China, the direct uses involve space heating, aquaculture and medicinal purposes. Algeria is using hot water for heating purpose only. In fact, the direct heat uses involve following things:

Table 8.2: Direct Heat uses of Energy from Earth

S. No.	*Applications*	*% used (approx.)*
1.	Industry	11%
2.	Heat pumps	11%
3.	Space heating	35%
4.	Bathing	13%
5.	Green houses	13%
6.	Aquaculture	12 %
7.	Agricultural drying	2%
8.	Snow melting air Conditioning	1%
9.	Others	2%

The reservoirs (present inside earth surface) contain hot steam and hot water. These are supplied to those surfaces, where they are used for direct use applications.

In fact, there are two main classifications, namely high temperature resources and low/moderate temperature resources. These low temperature resources are normally available on all continents and used for direct heating purposes or agriculture and aquaculture.

- **Advantages and Disadvantages of Geothermal Energy**

This energy possesses a number of advantages over conventional and other sources.

(*a*) **Advantages:**

1. This energy is least polluting compared to other conventional energy resources.
2. One of greatest advantages is its amenability for multiple uses from a single resource.
3. Since annual load factor (maximum) remains at 85% to 90% compared to 45% to 50% of other conventional resources.
4. Since zero fuel is used in this technology, so in this context, it is found to be cheaper compared to other energy sources.
5. This energy source delivers greater amount of net energy from its system than other alternative or conventional systems.
6. Geothermal energy is the source, which has practically no intermittency, has highest energy density and is versatile in its nature and so it gives greater benefits compared to other renewable energy sources.

(*b*) **Disadvantages:**

1. Drilling operation required in these sources is very much time consuming and noisy in operation.
2. Due to the withdrawal of large quantity of water/steam from a hydrothermal reservoir, surface subsidence may result.
3. Total net efficiency of this source is very low (≈ 12% to 15%) compared to other energy sources.

- **Limitations of Geothermal Energy Source**

Although the energy source has a number of advantages, but inspite of these, there exist a number of risks and limitations in the path of this source. One of the important hurdles is that after expanding effort on the exploration works, one may not be sure about the existence of heat resource at that location. Likewise drilling exploration works require a lot of money, which may not result in useful production sometimes.

Other limitation of this energy source is linked with the size of development and reservoir exhaustion. How a reservoir will work over a number of years , provides another risk in geothermal development. Subsequent assessment of resource size and production capacity is possible to reasonable levels of certainty and forms a critical part of any geothermal development.

2. TIDAL ENERGY (OCEAN ENERGY)

- **Introduction**

Tide is basically a periodic rise and fall of the water level of the sea. The energy obtained from the sea is called as tidal energy. This energy can accomplish a significant portion of all types of energies, which are renewable in nature. It is called renewable because the tides will continue to ebb and flow and the energy can be taken from it.

Actually the tide (in sea) can move a large quantity of water and by using this energy, a large amount of electrical power can be developed. Although the energy supply is reliable in nature and plentiful also, but converting it into electrical power involves extremely high capital costs. According to recent study, there exists an estimated potential of tidal energy of the order of approximately 8000 MW in the country. It includes 7000 MW in the Gulf of Cambay and nearly 1500 MW in the Gulf of Kutch in Gujarat. Also 100 MW potential exists in the Gangetic Delta in the Sunder bans region in West Bengal.

Fig. 8.7: Structure of tidal plant
[*Source*: BBC News, 19th Jan. 2011]

The working of this energy system is similar to a hydroelectric scheme, except that the dam is much bigger. Actually a huge dam (also called a barrage) is built across a river estuary. Due to the movement of tide (in and out), the water flows through the tunnels in the dam. Finally the ebb and flow of the tides can be utilized to turn the turbine-generator set, thus producing electricity.

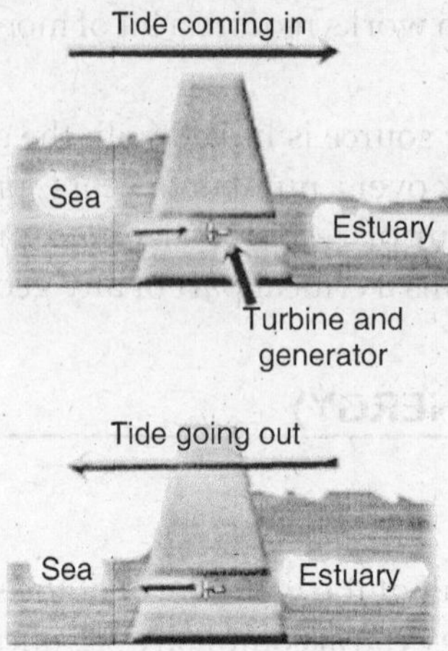

Fig. 8.8: Basic structure of tidal power plant

A major point of these tidal power stations is that the electrical power can be developed only, when the tide is either flowing in or going out (*i.e.* approx. 10 hours/day). Since we can know whether these tides will generate power at a particular time period or not, so we can plan to have other power stations generating at those times, when the tidal station is out of action. Another important issue is to use offshore turbines (rather like an underwater wind farm). It will definitely provide benefits of being much cheaper to build. It will also not have the problems related to environment that a tidal barrage would bring.

• Description of Various Basic Terms Related to Tidal Energy

In this case, we will try to focus on the brief summary related to tidal energy. We will discuss only the following terms:

(*i*) Ocean energy
(*ii*) Tidal power
(*iii*) Tidal fence and Tidal turbine
(*iv*) Wave power
(*v*) Conversion from wave energy to electricity through offshore and onshore systems

1. **Ocean energy:** More than 70% of the earth's surface is covered by oceans and the oceans produce mechanical energy from the tides and waves. One important fact regarding this ocean energy is that the sun affects all ocean activities. The gravitational pull of the moon primarily drives the tides and the wind powers the ocean waves.

   These ocean waves possess sufficient energy and forces that can be converted into electrical energy. Since ocean energy is non polluting as well as abundant, so researchers are exploring ways to make ocean energy economically competitive with fossil fuels and nuclear energy.

2. **Tidal power:** At those places, where tidal power generation is possible, the processing is similar to the hydroelectric power plants. Most important requirement regarding this thing is of barrage (or dam), which is made across a estuary or bay. Since the construction of these dams is very costly so the best tidal sites are those, where a bay has a narrow opening. When a proper level of difference in the water level creates, then gates are allowed to open. Thus water flows through the turbines and turbines drive the ac electric generator to produce electricity.

3. **Tidal fence and Tidal turbines:** Another important technology to harness tidal energy is the tidal fence. It actually appears like giant, turns tiles. These turn tiles spin via. tidal current, typical of coastal waters. Some of these currents run at 5 to 8 knots (*i.e.* 5.6 to 9 miles/hour) and produce larger amount of electrical power. Since the density of sea water is much more than air, so sea water possesses much more amount of energy compared to wind (air).

   Although at present, there is no tidal fence (in operation), but some countries are planning to construct it within few years from now. For example, a fence across the Dalupiri passage, between the islands of Dalpiri and samar in the Philippines, are in the manufacturing process.

   Likewise other technology is the tidal turbine. These look similar to wind turbines. At those places, where coastal currents run at the speed of nearly 4 and 5.5 mph, these turbines operate in best mode. Actually the best locations for tidal turbine farms are close to shore in water depths of 20 to 30 meters. Although at present, these farms are not in operation, but recently some suitable sites have been identified in Europe. Also some other countries, such as Japan, China and Indonesia have underwater turbine-farm sites that could be useful in future.

4. **Wave power:** From surface waves or from pressure fluctuations, energy can be extracted directly using wave power devices. These ocean waves can provide up to 2 terawatts of electricity. Although it may be a good source for electricity production, but wave power rich areas are very few in the world. These areas exist in South Africa, Australia, Northeastern and Northwestern coasts of United States, North Canada etc.

5. **Conversion from wave energy to electricity through offshore and onshore systems:** Wave energy can be converted into electricity through both offshore and onshore systems.

   (*i*) **Offshore systems:** These types of systems are situated deeply in the water. Typical bobbing mechanisms use the bobbing motion of the waves to power a pump that produces electricity. Some times offshore devices use hoses, connected to floats that ride the waves. The process takes place in the following manner.

   Rise and fall of float ⇒ Stretches and relaxes the hose ⇒ by which water is pressurized ⇒ Turbine rotates ⇒ electricity generates.

   In addition to this, seagoing vessels can also be used, so as to have the offshore waves energy. In this case, electrical energy is produced by funneling waves through internal turbines and then finally these are allowed to move back into the sea. These types of Vessels are now being used in (at present) Japan.

(*ii*) **Onshore systems.** These are those systems, which are built along shore lines to extract the energy from breaking waves. These (onshore systems) technologies are classified as follows:

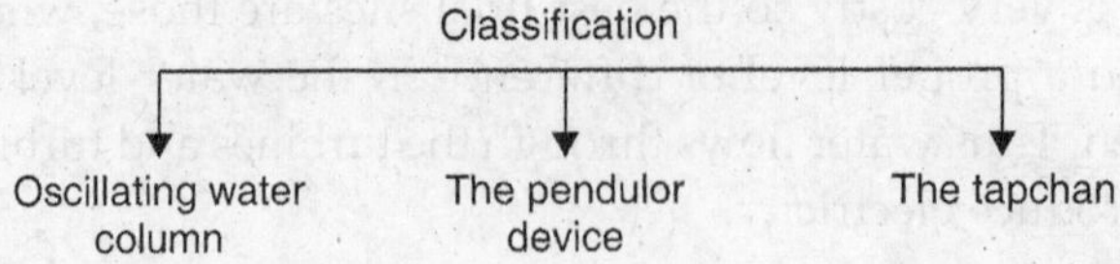

Fig. 8.9: On shore systems

(*a*) *Oscillating water column.* It consists of a steel structure, which is partially submerged and has an opening to the sea below the waterline. In encloses a column of air, above a column of water. As waves, enter the air column, water column rise and fall, due to which compression and depressurization of air column takes place. As the wave retreats, air is drawn back through the turbines. These types of water column devices have been built world wide. For example: On the Island of Pico in Azores, a 500 kW oscillating water column plant is being built. It will really provide sufficient electricity to several homes on Island. India also has such type of column wave-energy plant under testing at Vizhinjam in Kerala. Thus very soon, a no. of such plants will be established so as to take benefits from such kind of waves.

(*b*) *Pendular wave power device.* It actually consists of a rectangular box, which is open to the sea at one end. In such case, a flap is hinged over the opening. Flip swings back and forth, due to the wave power and thus the motion puts both the devices, hydraulic pump and generator, into operation. Worldwide, only few types of such systems are being utilized.

(*c*) *The tapchan.* This system is also called as tapered channel system, which is basically linked with hydroelectric power plant technology. The system consists of a tapered channel, which feeds into a reservoir constructed on cliffs above sea level. These waves spill over the walls of the channel into the reservoir and the stored water is fed through a turbine.

• Classification and Working of Tidal Power Plants

These tidal power plants are classified on the basis of the number of basins used for power generation.

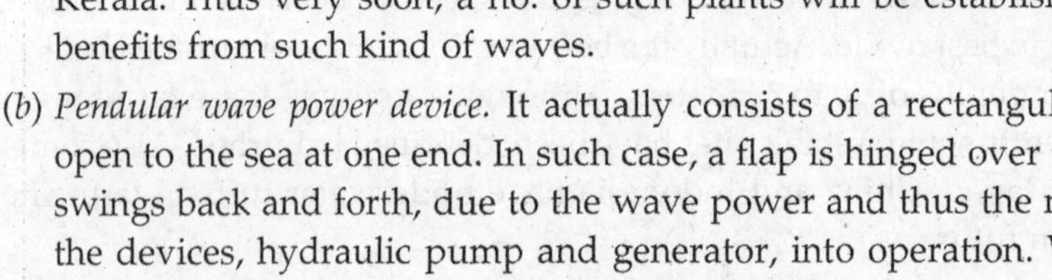

Fig. 8.10: Classification of Tidal Power Plants

Working. Working of various tidal power plants is as given below:

(1) *Single basin-one way cycle.* It is the simplest form of tidal power plant. In this system, during flood and ebb tide, the basin is filled. Here the water is allowed to flow, from the basin to the sea. Now it passes through turbine, which drives generator and power production takes place. One important point, which must be noted here, is that power is available, for a short duration during ebb tide.

(2) *Single basin-two way cycle:* It is another important type. In this, the generation period is increased, compared with one way cycle. The power is generated, during flood as well as ebb tide.

During the process, special turbine-generator sets are used for electricity generation. The main problem, with this arrangement is that the same turbine must be used as prime mover as ebb and tide flows, pass through the turbine in opposite directions.

(3) *Single basin-two way cycle with pump storage.* In this case, power is generated during both flood and ebb tides. For producing the difference in the water levels (between the basin and sea) at any time, a part of the energy is utilized. It is done by pumping water into the basin (up or down). Such types of plants exist in France. In these systems, power-production period is much larger compared to other techniques (as mentioned above).

(4) *Double basin type.* In this case, as the name suggests, two basins are used for power production process. One of the basins is intermittently filled by the flood tide and other is intermittently drained by the ebb tide.

One of the main important things here is that continuous power may be obtained with above system.

(5) *Double basin with pumping.* In such case, overall energy gain is possible, only if the pumping head is lower compared to basin-to-basin turbine generating head. OFF peak power from the base load plant, is utilized, so as to pump the water up the high basin.

- **Advantages and Disadvantages**

These tidal power plants possess both advantages and disadvantages.

Advantages:

1. This plant is not expensive and once have been built up , tidal power is free of cost.
2. Tides are totally predictable and produce electricity reliably.
3. No fuel is required and also no green house gases are produced.
4. Turbine namely, offshore turbines and vertical-axis turbines are cheaper and do not have large environmental impact.

Disadvantages:

1. In those situations, when tide is actually moving in or out, power is produced only at the rate of 10 hours each day.
2. A barrage (across an estuary) is very expensive to build.

3. MINI AND MICRO HYDROELECTRIC PLANT'S ENERGY

- **Introduction**

Hydroelectric power is generated by the flow of water through turbine (*i.e.* turning the blades of the turbine). A generator shaft connected to this turbine turns and hence generates electricity. Figure 8.11 shows the basic arrangement of small hydroelectric plant.

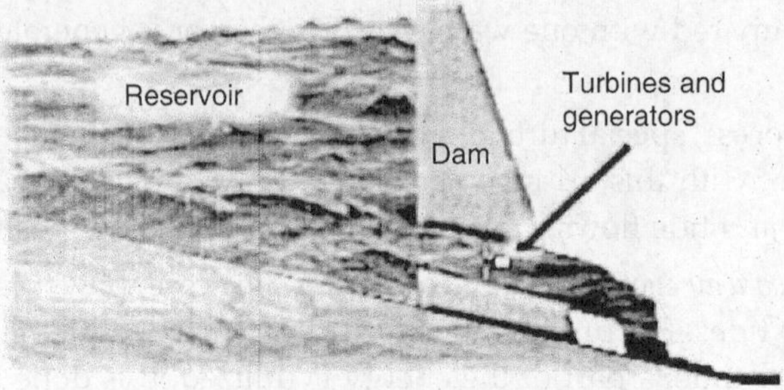

Fig, 8.11: Small hydroelectric plant arrangement

Nowadays due to the increased prices of oil, some new technologies are being proved as a centre of attraction. One of the important technologies is Mini and Micro hydro power plant technology. In reality, it is possible to set up a large number of small plants, known as micro and mini plants depending on their capacity. Hydropower accounts for nearly 10% of the world's commercial energy and for over 90% of the electricity generated in some countries like Norway, Kenya, Austria etc. Out of total power generation installed capacity in India, small hydro power contributes about 37367 Mw. It is expected that in near future, this power source will indicate a higher potential. If these small hydro power plants are set up all over the country, decentralized availability of power will be possible. Basically the concepts of small hydro vary from country to country depending on the size of its existing development. For example, in France and some other countries, only 3000 Kw is considered small hydro development, while in USA, 15000 Kw is considered. However in India, state wise numbers and aggregate capacity of small hydro power projects is as given in Table 8.3.

Table 8.3: State Wise Numbers and Aggregate Capacity of SHP Projects (Up to 25 MW)] [Installed & Under Implementation (As on 31-01-2011)]

S. No.	*State*	*Total Projected Installed Capacity (MW)*	*Projects Under Implementation With Capacity (MW)*
1	Andhra Pradesh	189.83	61.75
2	Arunachal Pradesh	78.835	38.71
3	Assam	27.11	15
4	Bihar	58.3	36.31
5	Chhattisgarh	19.05	1.2
6	Goa	0.05	—
7	Gujarat	12.6	

8	Haryana	70.1	3.4
9	Himachal Pradesh	375.385	132.2
10	Jammu & Kashmir	129.33	5.91
11	Jharkhand	4.05	34.85
12	Karnataka	725.05	107.5
13	Kerala	136.87	23.8
14	Madhya Pradesh	86.16	19.9
15	Maharashtra	263.825	51.7
16	Manipur	5.45	2.75
17	Meghalaya	31.03	1.7
18	Mizoram	36.47	0.5
19	Nagaland	28.67	4.2
20	Orissa	79.625	3.93
21	Punjab	153.2	21.4
22	Rajasthan	23.85	—
23	Sikkim	47.11	5.2
24	Tamilnadu	94.05	33
25	Tripura	16.01	—
26	Uttarpradesh	23.3	—
27	Uttrakhand	134.12	230.65
28	West Bengal	98.9	79.25
29	A&N Islands	5.25	—

In addition to this, there exist several small hydro power projects with private sector, which are as given [with their capacity (MW)] in Table 8.4.

Table 8.4: Small Hydro Power Projects (With Private Sector) [As on 31-12-2010]

S. No.	*State*	*Total Number*	*Total Capacity (MW)*
1	Karnataka	95	694.90
2	Himachal Pradesh	63	271.25
3	Andhra Pradesh	43	104.43
4	Punjab	18	26.20
5	Maharashtra	13	74.00
6	Uttaranchal	10	48.30
7	West Bengal	5	6.45
8	Kerala	3	36.00

9	Gujarat	2	5.6
10	Haryana	2	7.4
11	Jammu & Kashmir	2	17.5
12	Orissa	2	32.00
13	Assam	1	0.10
14	Madhya Pradesh	1	2.20
15	Tamil Nadu	1	0.35

Actually small hydro development can be classified into two types:

1. (*a*) Small discharges but having high head.
   (*b*)Physical size of plant required is small.
2. (*a*) Large discharges under very small head.
   (*b*) Physical size is big compared to general notions of the rating.

Classification of Small Hydro Power Stations

Hydel power plants can be classified, as according to their capacity and their heads. Now let's have a brief summary regarding these two, as given in Table 8.5 and 8.6.

Table 8.5: Depending Upon Capacity

S. No.	*Category*	*Capacity*	*Application*
1.	Large hydel plant	50 MW to 1000 MW	Large cities
2.	Small hydel plant	1 MW to 50 MW	Small cities to town
3.	Mini hydel plant	100 kW to 1000 kW	Towns
4.	Micro hydel plant	< 100 kW	Rural community
5.	Pico hydel plant	< 5 kW	Individual home

Table 8.6: Depending on Head

S. No.	*Type of head*	*Depth*
1.	Ultra low head	Below 3 meter
2.	Low head	Less than 30 meters
3.	Medium head	Between 30 to 75 meters
4.	High head	Above 75 meters

(* Above mentioned data may vary from country to country)

⇓

Now on the basis of heads available, small hydel schemes (mini, micro and small) may be categorized as:

(*i*) Independent scheme (*ii*) Subordinate scheme

- *Independent Scheme.* In these schemes, medium head scheme is generally used or preferred compared to low and high head. Basically these are the schemes, where a stream flow is captured, regulated and developed for the principle objective of power generation only.

These schemes are successful in Jammu and Kashmir, Manipur, Nagaland, Himachal Pradesh etc.

- *Subordinate scheme.* In these schemes, primary aim is irrigation or drinking water and secondary aim is power generation. These regions, in which water is stored in the form of reservoir, are utilized for subordinate scheme. These schemes are preferred in peninsular India.

- **Components of Hydroelectric Scheme**

The basic and common components of a hydroelectric scheme is as given below:

(*a*) Diversion and intake structure
(*b*) Densilting Chamber
(*c*) Water conductor system
(*d*) Forebay/ balancing reservoir
(*e*) Surge tank
(*f*) Pen stock
(*g*) Spill way
(*h*) Power house building

(*a*) **Diversion and intake structure.** This structure must be simple and economical. This structure may be of two types, such as solid boulder structure and trench type weir structure. First type of structure could be used, if boulders are available in the river and rock is encountered in the river bed within one meter depth. One important benefit is that these boulders weirs are quite cheaper. Latter type is suitable, where rock is not available in river bed. In reality, it consists of a trapezoidal trough, which is located below the bed of river.

(*b*) **Densilting chamber.** It is a chamber, which is necessary at those locations, where the water contains large quantities of course silt to reduce erosion damage to the turbine runner etc. The silt particle's size varies from 0.2 to 0.5 mm, for medium size hydel power stations and from 0.1 to 0.2 mm, for high head. Also the depth of tank may be kept between 1.5 m – 4 m.

(*c*) **Water conductor system.** It is another important component of hydropower scheme. This system must be designed in such a way, so that there is least loss of head and water, due to seepage. Also velocity of flow must be adequate, so as to prevent reduction of discharging capacity, due to setting of silt.

(*d*) **Forebay/ balancing reservoir.** Forebay may be used as a balancing reservoir. About 4–6 hours could be provided for storage purpose. However if the forebay is just used as a transit point, a storage of about 2 minutes may be sufficient.

(*e*) **Surge tank.** This tank is of great importance, when the length of water conductor conduit is more than 5 times the load on the machines. Sometimes, it may not be provided, where the pressure surges may be avoided with deflection of jet.

(*f*) **Penstock.** Water is fed to the machines through penstocks. These can be made up of steel pipes, Hume pipes and PVC pipes as according to the design pressure. For continuous and smooth entry of water from forebay to the penstock, bell mouth entry is provided. Penstock may be designed into two ways, namely structural design and hydraulic design. In former case, penstock's thickness and in latter case, Penstock's diameter is obtained.

(*g*) **Spill way.** It is another portion, which is provided, keeping it crest at the permissible water level, so that the level of water may not rise, above the maximum permissible level.

(*h*) **Power house building.** This building consists of all the parts, such as turbine generator, control panels, auxiliary equipments, which makes the process (of small hydel power plant) simpler. This building must have a strong base and simple structure in R.C.C. Its size depends upon the total number of machineries required and the distance between them. Also the building height is kept from 3–5 meter.

In small hydropower stations, (SHP) controls are required for unit start, unit shut down, unit synchronizing and unit loading, unit electrical and mechanical protection, emergency shut down etc. The major cost in these systems is covered in operation and management sections. If mini and micro plants, require synchronizing and starting facilities, then obviously a precise control will be required and major cost will be involved in its operation.

• Advantages and Limitations of Small Scale Hydroelectric Plants

Advantages:

1. Once these power stations are built, the running expenditure becomes very low.
2. These SHP have low gestation period *i.e.* from 8 to 24 months.
3. SHP releases no heat and no pollution, so adverse environmental impacts are negligible and for small installations, they may be completely eliminated.
4. Their operation is in hilly and remote areas, so they possess lent added attraction.
5. Since these plants are inter connected and synchronized with grid, plant utilization factor, becomes very high, so SHP becomes economically attractive.
6. Investment/kW installed is not very high due to the development of compact efficient machines, so SHP projects do not require large capital investment.
7. In India, most of these schemes are being constructed on existing canals and irrigation system, so there exists minimal disturbances to surrounding environment.

Limitations:

1. Sites are in hilly and remote locations and at those locations, geological conditions are not suitable.
2. Prohibitive economics compared to that of conventional hydro power, which is still abundantly available in the country.
3. Lack of awareness of benefits from small development.

- **Hybrid Systems**

SHPs are used to meet mainly rural energy requirements. These plants may be used with other sources so as to make a hybrid-system, to meet out the requirements of power, for irrigation, street lightening, domestic purpose etc. In these hybrid systems, micro hydel system may be used as base load plant and biomass-gasification based electrical generation system as peak load plant. One of the hybrid systems is in operation at Kakroi (Haryana), where a combination of 300 kW of micro-hydel and 200 kW of biomass gasification based plant is being used. In brief, it can be said that the hybrid system is capable to meet out entire-energy requirement, if suitable materials as well as locations are easily available.

4. FUTURE POWER GENERATION SOURCES

(*a*) **Hydrogen as the wonder fuel**

(*b*) **HELIUM–3 as a source of energy**

- **Sources of Power Generation in Future**

(*a*) Hydrogen as the wonder-fuel

- **Introduction**

  Hydrogen is the simplest element known to man. Each atom of hydrogen has only one proton and one electron. Our Sun is just like a ball made up of both hydrogen and helium-gases. Actually inside the Sun, hydrogen atoms combine to form helium atoms. This overall process is called fusion process, which gives off radiant energy. This energy provides us light and makes the wind blow, rain fall and plants grow.

  Actually Hydrogen (H_2) doesn't exist on earth. It is always mixed with other elements. For example, when it is combined with oxygen, it makes water. Likewise, when it is combined with carbon, various compounds, such as CH_4, coal and petroleum are made. The basic structure of hydrogen atom is as given in Fig. 8.12.

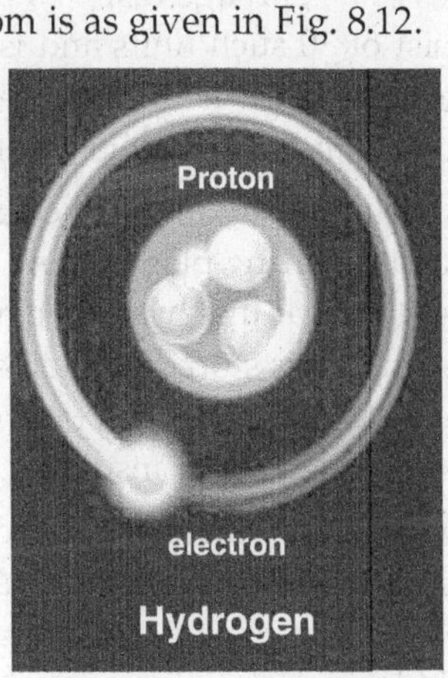

Fig. 8.12: Basic structure of hydrogen atom

Hydrogen can be easily separated from hydrocarbons by a process, known as "reforming". Currently most of the hydrogen is made this way from natural gas. In reality, hydrogen possesses a large amount of energy within itself. By using hydrogen, as a fuel, very less amount of pollution will be created in atmosphere. Even NASA has used hydrogen (in liquid form) to propel the rockets and space shuttle into orbit. Also, the fuel cells powered by hydrogen, maintain the shuttle's electrical systems, thus producing clean by product. [Many energy sources can be used to produce hydrogen. Hydrogen can store the energy until it is needed and move it to where it is needed.]

- **Hydrogen as Fuel**

It is the most abundant of various chemical elements. Hydrogen is a gas at normal temperature and pressure. In general, it may be used as a fuel in I.C. engines, in fuel cells so as to produce heat and electricity etc. It is that kind of gas, which cannot be easily captured for direct use purpose. The reason behind this is that it is so light, that free hydrogen gas tends to drift up and away from the surface of earth.

Hydrogen gas can be produced from a number of chemical compounds, containing hydrogen. As we all know that H_2O is the most well known and abundant source of hydrogen. In this case, by using electrolysis process, H_2 can be obtained. The produced hydrogen is then collected and stored for future use. Hydrogen can also be obtained from natural gas, which is primarily methane (CH_4). By using a number of chemical processes (reactions), hydrogen molecules can be splitted (with the help of reforming of methane). Again, the hydrogen is collected and stored for use. Also by using alcohols (methanol, ethanol etc.) and hydrocarbons (gasoline), H_2 can be produced.

Nowadays, a new report about hydrogen is that Garbage and sunlight can make hydrogen. From the reliable sources, it has been known that company called "Solar Hydrogen Energy Corporation" (SHEC) based in Canada, has devised a novel technology so as to produce hydrogen from garbage using sunlight. The planning within few years, is to erect arrays of solar powered units, by which local land fill methane would be converted into hydrogen. This project, when implemented would not only provide a cheap source of hydrogen but also rid the civic authorities of its garbage disposal problems. The proposed installation at the land fill would consist of 30 such units and is estimated to reduce nearly 25% of city's greenhouse gas emissions. Thus it is hoped that in near future, the arrays would create more than 2.6 million pounds of hydrogen per year.

Now the question arises that how would it be used? Actually hydrogen could be produced at centralized manufacturing plants. Now by using pipelines or tankers, it can be distributed to end users. It is clear that hydrogen must be in the form of either a gas or liquid. If it is used as a gas, then it is stored in a tank, at slightly higher pressure, due to lower energy density. Also it can be stored in a tank containing metal hybride. Likewise if it is used as liquid, it is stored in a super insulated tank at very low temperatures.

At present, the lack of hydrogen manufacturing and distribution infrastructure is a major roadblock to moving beyond demonstration projects to a wider use of hydrogen by the public and other users. In future, we can hope that a solution to this problem will exist and it will remove all road blocks.

- **Hydrogen Energy**

In today' world, large scale industries are powering out gases (like CO_2, NO_x, SO_x etc.) and sludge. Due to these pollutants, rivers and water bodies are now becoming a storage media for industrial waste and garbage. However if we look at India's situation, then we may find that, there exists a further problem of high population, causing uncontrolled pollution (*i.e.* due to various human activities). In villages, a large fraction of population lives and their contribution to present day pollution of air, land and water is not negligible. We use coal and petroleum so as to meet out our daily needs, but it is very inefficient way of using the energy due to the reason that most of its part is wasted during the process of conversion in the power plants. Also the efficiency of the above process remains nearly equal to 35%. In this context, hydrogen gas, as a source of energy offers definite advantages. It is not a pollutating fuel, so reduces greenhouse gases effect. The demand for H_2 is also based on its use as raw materials for fertilizer and plastics as a reducing gas for the manufacture of iron sponge, for manufacture of synthetic fuels like hydrocarbons, methanol, SNG and in many other commercially important products. So in this context of the global situation, it can be said that fossil fuel exhaustion is taking place at a very large level. That's why some other alternative sources must be utilized so as to get electrical energy and to reduce the atmospheric pollution. H_2 gas has been proved as one of the sources, that is non polluting and can be obtained from the abundantly available raw materials (*i.e.* water (H_2O) by electrolysis or photolysis). As hydrogen is not a primary source of energy but is rather a secondary source of electricity, so convenient and environmentally friendly ways of storage and transportation are required. As discussed in the previous section, it is clear that hydrogen may be stored as a compressed gas, metal hybride or as a cryogenic liquid. Since H_2 gas possesses large weight as well as volume, so gaseous storage is not feasible and liquefication process is comparatively expensive. Out of these, the best method is storage as metal hydride. In these compounds, hydrogen density per unit volume is high.

However if Solar cells (not expensive) are easily available, then energy can be stored in the form of hydrogen. Excess amount of electric power generated during the day may be utilized so as to electrolyze water to produce H_2 and O_2 gas etc. (which can be compressed in storage tanks and used later in H_2-O_2 fuel cells for electricity generation). In the electrical network, if the electric power goes to peak load, H_2 can be converted to electrical energy using fuel cells. On the other hand, excess electrical energy can be stored as hydrogen energy via electrolysis process. There is not doubt, that considerable amount of potential for hydrogen energy exists in our country and serious and sincere efforts in this direction are desirable. If efforts are made in this direction, then H_2 will become a major fuel, replacing established sources of energy. Thus India has all the potentials to be an active partner but safety considerations should be kept in mind.

- **Uses of Hydrogen**

Hydrogen is used in industrial applications, such as refining treating metals and food processing. It is also being used as a fuel for "zero emissions" vehicles, to heat homes and offices to produce electricity and to fuel aircraft. This hydrogen can be mixed with methanol, ethanol, gasoline etc., to increase performance and reduce pollution. It is an amazing fact, that Nitrogen oxide (NO_x) emissions can be reduced by adding nearly 5% hydrogen to gasoline.

In the present scenario, the greatest use of hydrogen is in the NASA space programmes. This is the only fuel, which has propelled the space shuttle and other rockets since the 1970's. These space shuttle makes use of fuel cell (hydrogen mixed) to run its computer systems. The fuel cells, reverse electrolysis process—Hydrogen and oxygen are mixed so as to produce electricity. Hydrogen has also been proved as a perfect carrier for energy. It can be used to store and distribute the energy, whenever it is needed. Besides various other uses, it can be used in large quantities in petrochemical industry and for reduction of iron oxides in the steel industry. Hydrogen can be produced, stored and used, as safely as other fuels. The goal of India must be for hydrogen to produce 10% of our total energy demand by the year 2030. It may reduce our dependence on fossil-fuel and provide us cleaner source of energy in near future.

(*b*) Helium-3 as a source of energy

- **Introduction**

It is that energy source, which could provide power for the future. We expect that by the year 2050, population of world will cross the limit of 10 billion. In that situation, no fossil fuel, coal will exist on earth and we will have no alternative resources but fusion energy. In this regard, Helium-3 is a completely clean source of energy and the basic arrangement of helium may easily be represented by Fig. 8.13.

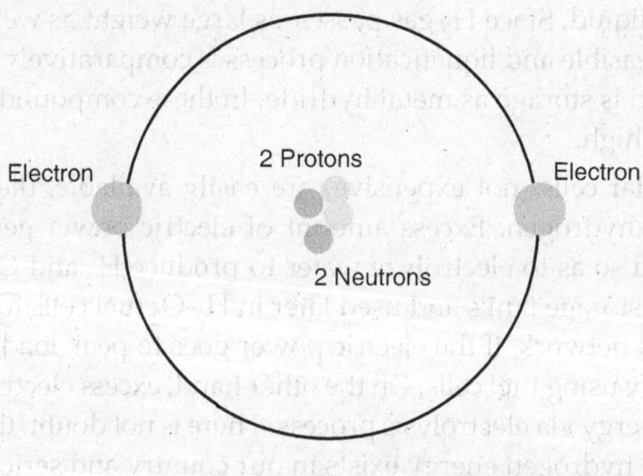

Fig. 8.13: Basic arrangement of helium atom

Two helium-3 atoms may be fused in a thermo nuclear reactor so as to produce normal helium and energy. It becomes a good energy source, due to the reason that this fuel is non radioactive, so the process and the residues produce no radioactivity. But the major point is that this reaction takes place at 10 times the temperature of the TOKAMAK (a reactor).

Actually very little amount of helium-3 exists on earth and it is from the time of formation of earth itself. However the moon has much more quantity of helium-3 compared to earth. This source of energy is really a wonderful source, because it has manifold advantages. Nearly 25 tonnes of helium-3 have been used in U.S., for a whole year in 1993.

- **Lunar Helium-3 as an Energy Source**

In the reaction for nuclear fusion, main ingredients used are tritium, deuterium, isotopes of helium etc. We know that Neutrons and protons form nucleus of atom. The term is called as nuclear reaction, because in this review, we are concerned with rearranging the atom's nucleus. In general, neutrons (neutrally charged) keep the positively charged protons from fighting each other. Hydrogen is important material because it is used to produce water by combining it with oxygen. (H_2 atom possesses 1 proton and no neutrons). Likewise, Deuterium is an isotope of hydrogen that has a neutron next to its lonely proton.

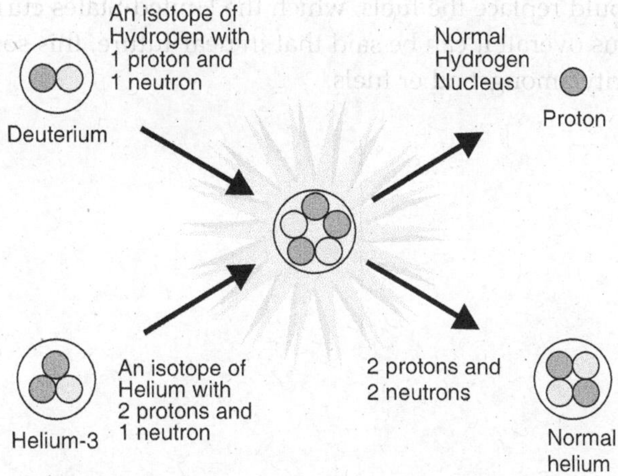

Fig. 8.14: Reaction process for helium- 3

Now, the most important case is of helium gas. Normal helium has 2 protons and 2 neutrons in its nucleus, thereby giving it an atomic weight of 4. At this stage, if one of the neutrons is kicked out, helium-3 is obtained. The real fact is that this happens once in a while in very energetic nuclear reactors, especially the sun. Helium-3 may grab neutron in the deuterium, if it is given a chance. Obviously it is given a chance by introducing the Helium-3 to the deuterium at a high temperature.

Helium-3 is mixed with deuterium to produce enormous amount of energy. Overall reaction is as follows:

$$D + {}^{3}He \rightarrow P\ (14.7\ Mev.) + {}^{4}He\ (3.7\ Mev.) + 18.4\ Mev.$$

It is basically a nuclear fusion reaction. In this reaction, helium-3 and deuterium atoms join each other and give off a proton and helium-4. Here the output product weight less compared to initial compounds, because the missing mass converts into energy. Great thing regarding fusion reactor is that A reactor built to use the D-He-3 reaction would be safe and there will be no significant exposure to radiation.

- **The Supply and Energy**

  He-3 is basically a by-product of the maintenance of nuclear weapons, which would supply us with about 300 kg of He-3 and could continue to produce about 15 kg per year. A recent report on He-3 shows that a large amount of it has been deposited by the solar wind in the lunar regolith. Thus it is obvious that we shall find He-3 down to depths of several meters. Actually the highest concentrations are in the lunar maria, about ½ the He-3 is deposited in the 20% of the lunar surface covered by the maria.

  Similarly in continuation to this, it can be said that nearly 20,000 terawatt-hours of thermal energy would be produced, when 1 million metric tonnes of He-3, is reacted with deuterium. It means about 10 times the energy, which could be achieved from mining all the fossil fuels on our earth, without the smog and acid rain. For example in U.S.A., about 25 tonnes of He-3 would be used for power supply and generation purpose. In brief, He-3 would replace the fuels, which the United-States currently buys to generate electricity. Thus overall it can be said that in near future, this source will be useful and attain popularity amongst other fuels.

Chapter 9

ENERGY CONSERVATION AND CLEANER ENERGY PERSPECTIVE

ENERGY CONSERVATION

Introduction

Energy conservation is the wise and effective use of energy so as to ensure that for a given amount of energy, maximum action, productive work and profitability are obtained. Here the word "**Conservation**" demands efficient, economical and careful use of energy. As according to the words of V.S. Verma [DG–BEE], the definition is as follows:

"Energy conservation is an objective to which all citizens in the country can contribute, whether a household or a factory, a small shop or a large commercial building, a farmer or an office worker, energy user and producer of energy can and must make this effort for his own benefit, as well as that nation".

There exists many ways, by which energy may be conserved. These methods include, (*i*) No cost/low cost method (*ii*) Medium cost method (*iii*) High cost method. Out of all these methods, low cost method requires least effort.

Energy Conservation—a Growing Concern

As we know that energy resources are finite so immediate steps must be taken to get energy conservation. Actually energy production in India has not been able to keep pace with consumption; its efficient use is becoming increasingly important. One unit of electricity saved reduces one kg coal consumption, which is a non-renewable resource. That's why immediate steps need to be initiated to get into energy conservation. Thus, India is now continuously working on a comprehensive package to popularize it (energy conservation). Energy conservation received great attention in India in the 70's. Actually awareness about the advantages of energy conservation exists due to the rapid increase in energy demand. Structural changes in the economy have led to the expansion of the industrial base in the country and a renewed effort on energy conservation.

Due to increase in the prices of oil, India will have to move towards alternative sources of energy, like gas and coal bed methane so as to meet its energy requirement. Steps must also be taken to initiate action to reduce the energy intensity of various consuming sectors of economy and promote conservation and demand management through appropriate organizational and fiscal measures. Measures must be introduced so as to meet the basic energy needs of the rural and urban households and to reduce the existing inequities in energy use and also to maximize satisfaction of demand for energy from indigenous resources.

Energy conservation must be motivated by various sectors. The motivation for energy conservation depends on the sector, in which it is to be used. These motivations for energy conservation in Government sector may be as follows:

(*a*) to maintain proper energy supplies

(*b*) to satisfy energy saving goals etc.

Likewise for industrial sector, the motivations for energy conservation may be:

(*a*) to ensure adequate energy supply required for the production processes.

(*b*) to maintain profits of the unit in the face of increasing energy costs.

Due to the implementation of energy conservation plans, various advantages may be as follows:

(*i*) lesser maintenance cost

(*ii*) labour cost saving

(*iii*) reduced overall cost of heating and cooling appliances

(*iv*) smaller size of equipment

(*v*) fuel savings etc.

For realizing the importance of energy conservation, the Vijayawada Local Centre in India of the Institution has taken the task of awakening and propogating the need for energy efficiency and conservation amongst people by forming the Energy Conservation Mission cell.

ENERGY AUDIT

Energy auditing is one of the most important tools for energy conservation and for achieving energy efficiency. Now let's understand the basic meaning of "energy" and "audit" and then we will be able to adopt energy auditing.

Actually energy is the ability to do work. It is basically the capacity of matter or radiation to do physical work. Likewise an audit is an examination of accounts or dealing with money or property by proper officers or persons appointed for that purpose. Thus in this sense, "**energy audit**" may be defined as follows:

An energy audit is an in-depth study of a facility to determine how and where energy is being used or converted from one form to another, to identify opportunities so as to reduce energy usage to evaluate the economics and technical practicability of implementing these reductions and to formulate prioritized recommendations for implementing process improvements to save energy. An energy audit will help the management to:

(*a*) How much in the furnace, in the boiler, in air conditioning, in the main plant and other utilities etc. A prerequisite for energy audits is to quantify energy distribution and wherever possible, to relate to production or other activities.

(*b*) Understand more about the way, energy is used in the plant.

(*c*) An audit should attempt to balance the total input of energy with use and cost. It represents, how much energy came into the business during a month and how it was used in a plant.

Energy Audit's Types

These can be classified into two main categories:

1. Preliminary audit   2. Detailed audit

1. **Preliminary audit:** It is carried out in a limited time period *i.e.* from 1 to 10 days. This audit mainly tells us about the cost of energy and wastages in the major equipments and processes. It also gives the major energy supplies and demanding accounting.
2. **Detailed audit:** The time period may be from 1 to 10 weeks. It accounts for the total energy utilized in plants and involves detailed engineering for options so as to reduce cost/consumption.

However for the achievement of energy conservation, using energy audit, three different phases may be used and these include:

(*a*) short term   (*b*) medium term   (*c*) long term.

CLEANER ENERGY-CLEANER FUEL–ENVIRONMENTAL CONSIDERATIONS

• Introduction

Clean power is that power, which is tapped by renewable technologies, such as solar, wind, biomass etc. These technologies produce no pollution (very little), so these may be the power of future. Now the question may arise that why we prefer cleaner energy? The answer may be given as follows:

It is really good for all of us because this energy benefits every one. Developing energy and power by cleaner technologies instead of power generated by fossil fuel, protects the fragile environment, we all need to live and be healthy. We know that when we use fossil fuels, pollution is created. These pollutants include CO_2 and nitro sulphur dioxide. These are dangerous substances, which pollute the air and enter into our bodies. It means that serious consequences will take place for all living creatures. In this context, clean energy must be preferred, (which may be obtained by renewable technologies) which makes us healthier.

• Carbon Intensity and Renewables

India has a strong environmental movement. It is continuously focusing on the conservation of forests and wild life, so as to prevent environmental (vehicular and industrial) pollution. India realizes that conventional energy sources are the biggest polluter and over 60% of greenhouse gas emissions is caused by burning coal alone. In this context, it has strongly come out in support of development of renewables. These are the renewable sources alone, which can reduce the carbon intensity in the atmosphere.

Now let's take a look at India's situation regarding carbon intensity. India's carbon intensity has remained relatively flat over the past 20 years, even falling from a high in 1995, back below 1980 level. India's still elevated energy intensity level is due in large part of its economic expansion, coupled with the virtual absence of energy efficiency and conservation measure in most industrial sectors. India mainly depends upon coal of low quality, so it goes a long way towards explaining the country's relatively high carbon intensity level. Indian economic policies such as high import tariffs on high quality coal and subsidies on low quality domestic also have contributed to increased use of low quality coal. Although initiatives to encourage the use of higher quality coal, such as reducing the tariff on imported coal, may reduce the country's carbon intensity.

Actually electricity is one of the keys of both India's economic development and its environmental protection efforts. Some firms use less efficient and older techniques to produce electricity, so in this context, India has initiated incentive-schemes for producing electricity from efficient ways so as to discourage these kind of firms.

Indian government has also used fiscal incentives, in the form of customs waivers and soft loans, to encourage the installation of pollution abatement equipment market mechanisms such as user charges, deposit refund systems, marketable permits and taxes for implementing pollution measures. India has made significant efforts in the field of environmental protection, developing environmental standards for both products and processes, requiring environmental impact statements in certain areas and introducing environmental audits. Actually the Indian government's ability to safeguard the country's environment will depend on its success in promoting policies that keep the economy growing while providing adequate energy needs to satisfy the population's growing energy consumption requirements in a sustainable manner.

- **Clean Energy Programmes**

(*a*) Clean Air initiative—Asia programme

This CAI-Asia programme is a programme that includes following goals:

- to improve regulatory frameworks at regional level
- to share various experiences on air-quality management, etc.

Actually for improving the quality of air in urban cities, technical knowledge as well as experience, both are desired and both these things are available in cost effective ways. The examples of CAI-Asia tasks could be:

- promoting the use of cleaner fuels and vehicles, public private partnership's, awareness of impacts and solutions, the planning and operation of public transport etc.
- advancing the introduction of cleaner technologies for energy generation and industrial production.

To help Asian cities, The Asian Development Bank (ADB) has approved a great amount for technical assistance. In this case, air pollution will be reduced by promoting the main streaming of air quality management in policies and programmes of its developing member countries. Thus CAI (for Asian cities), will help in implementing the project. In other words, it can be said that the main aim of the project will be to improve the quality of air (stabilize air pollution level). Another important aim of this project is the mainstreaming of AQM issues within ADB by developing a pipeline of AQM projects for ADB and other development institutions.

(*b*) **Connecticut clean energy communities programme.** The Connecticut clean energy communities programme is a partnership between the Connecticut clean energy fund and smart power, whose purpose is to assist communities in the purchase and support of clean energy. This type of programme has been designed to provide commitment with opportunities to purchase and support clean energy at no cost, by making the purchase of clean energy, a viable and attractive option for both business and residents. Due to this programme, we may have large benefits, which include the following:

It reduces emission of greenhouse gases that cause global warming. In addition to this, it also reduces electricity demand on the state's transmission and distribution system through the installation of on site clean energy system.

- **LPG—Reduction in CO_2 and Greenhouse Gases**

LPG is a pure, clean energy source, which may be used for a wide range of industrial uses. It is not just useful for industrial processes but can be a vital component in civil engineering, being used for everything from heating bitumen for road building to plumbing and brick manufacture. It (LPG) may be proved helpful in reducing harmful exhaust emissions, caused by traffic and due to its low carbon content, LPG engine reduce the emission of CO_2, the main greenhouse gas. Recent tests have shown the following advantages of LPG based engine emission:

(*i*) **Compared with ultra low sulphur petrol**

11% to 13% less carbon dioxide

15% to 80% fewer oxides of nitrogen

20% to 40% fewer hydrocarbons

30% to 35% less carbon monoxide

(*ii*) **Compared with ultra low sulphur diesel**

80% to 95% fewer particulates

99% to 99.8% fewer ultrafine particles

90% to 99% fewer oxides of nitrogen

Thus in brief, it can be said that LPG has the potential to reduce emissions from vehicle exhausts, below those of conventional fuels.

- **Solar Thermal Reduces GHG (Greenhouse Gas) Emissions in an Efficient Manner—Says Study**

Recently a report by Green Markets International tells us that solar water heating may be more effective at reducing GHG emissions. According to this report, solar water heating contributes to economic development in a number of ways, such as low cost manufacturing equipment and creation of new job opportunities in manufacturing, retail sales, business administration, system design, installation and maintenance of solar water heating technology. One fact regarding GHG emissions is that, solar water heating is even more cost effective in reducing GHG emissions, when the equipment is locally manufactured and the carbon intensity of the baseline fuel is high. The report also points out that despite the potential environmental and economic benefits of wide-spread solar water heating, a multitude of barriers still hinder the technology's broader application, which include insufficient quality control and lack of awareness. Finally as a conclusion, report established that solar water heating can contribute substantially to carbon abatement, while supporting the achievement of economic development goals and carbon finance can help overcome barriers to the broader adoption of solar water heating technology.

- **Cleaner Production—Clean Technologies**

In today's scenario, the fallout has been large scale propogation of obsolete technologies, which have impacted our environment and we can see today excessive wastage of natural resources and contamination of air water and land. Actually the "Policy Statement of pollution Abatement" since 1992 is providing impetus to this change from pollution control to pollution prevention through attitudinal shifts policy reorientation and green credit facilities etc. Now let's discuss briefly about all these things:

- **Attitudinal Shifts**

(*a*) Emphasis is given on overall management of quality using maximum waste reduction.

(*b*) Entrepreneurs are motivated to adopt cleaner production/waste, minimization as means of improving economic returns.

- **Policy Reorientation**

(*a*) To discourage import to obsolete technologies.

(*b*) To start incentive schemes for development of environmentally sound technologies.

- **Greener Credit**

(*a*) To provide soft loans for switching over from polluting technologies to ESTs.

(*b*) To have abolition of subsidies on energy and raw materials in the existing units depending upon status of technology.

- **Clean Energy on a new-high**

2007 was proved as a successful year for clean energy. According to a report **"clean energy trends 2008"**, (U.S.) there will be no slow down in next year.

Some trends for growth, projected by the report up to 2017 are as follows:

(*a*) Biofuels reached at $ 25.4 billion in year 2007 and is expected to reach $ 81.1 billion by 2017.

(*b*) Solar PVs is expected to reach at a level of $ 74 billion by 2017.

(c) Likewise wind power and fuel cell market will reach up to $ 83.4 billion and $ 16 billion respectively in 2017.

Some of the key trends are as follows:

- **Rise of Sustainable Cities**

Some places in various countries are planning to develop clean technology policies. Abu Dhabi (in United Arab Emirates) is in the process of developing the world's first zero carbon, zero waste, car free city, called 'Masdar'. Likewise Mayors of more than 500 cities in U.S. have reduced their GHGs to Kyoto protocol targets.

- **The Rise of Geothermal Power**

Due to improved technology, geothermal power is emerging as a clean energy source, especially in USA. Some new projects in 12 states will, double the US geothermal capacity and it will raise to nearly the 6300 MW, enough to power 6 million homes. It is hoped that by the year 2050, geothermal power will reach to a potential of 100 GW, thus overall it can be said that in the near future, clean energy may reach on a peak point.

BIBLIOGRAPHY

REFERENCES

1. Rai G.D., Non-Conventional Energy Sources, Khanna Pub. 4th Ed., 2000.
2. Gupta B.R., Energy Conservation by All, Symposium on Energy Conservation, IE (India), 1975.
3. Gupta J.B., A Text-Book of Power System, S.K. Kataria & Sons, 2000.
4. Gupta B.R., Generation of Electrical Energy, Eurasion Publishing House, 1998.
5. Annual Reports by Ministry of Non-Conventional Energy Sources (MNES), Govt. of India, 2003-04.
6. Ranga Rajan S., Policy for Wind Energy and its Utilization in India, 1995.
7. Energy Studies Yearbook: 1993 (New York—United Nations), 1995.
8. Khare P.N., MSEB, Article on Power Theft—A Root Cause of T&D losses.
9. World development report, 1997.
10. International energy agency, 1998 and 1999.
11. Green energy world magazine, Vol. 2, No. 1, Jan.-Feb. 2006.
12. E-power magazine, May-June, 2005.
13. IEI news, Volume 55 (N), May, 2005.
14. K. Katri Pradeep and K. Khedkar Mohan, Electrical India Magazine, November, 2005.
15. CII–Godrej, GBC Publication, RES–Factsheet, No. 2, June, 2004.
16. Green energy world magazine, Vol. 2, No. 3, May-June, 2005.

Internet Site For Supplementary Reading Material

1. http://www.mnes.nic.in
2. http://www.eia.doe.gov/
3. http://www.nrel.gov/
4. http://www.renewableenergy/com/
5. http://www.renewableenergyaccess.com/rea
6. http://www.energy.com.au/switched-on/activites/photovoltaic.htm
7. http://www.indiastat.com/india/showdata
8. http://www.howstuffworks.com/fuel-cell.html
9. http://www.eere.energy.gov/biomass
10. http://www.windpower-monthly.com
11. http://solstice.crest.org/renewables/ireda/wind.html
12. http://www.asi.org/adb/02/09/he3-intro.html

INDEX